This book is a record of cultural itinerary between the two cities, where each city one leaves is not the same as that to which one returns.

Este libroesunregistro de un itinerario cultural entre dos ciudades, dondecada ciudad queunodeja no es la misma a la queunoregresa.

이 책은 두 도시에 대한 여행의 기록이다.
이 여행에서 다시 돌아온 자의 도시는 떠날 때의 도시가 아닌 다른 곳이 되어 있을 것이다.

This book is a record of cultural itinerary between the two cities, where each city one leaves is not the same as that to which one returns.

Este libroesunregistro de un itinerario cultural entre dos ciudades, dondecada ciudad queunodeja no es la misma a la queunoregresa.

이 책은 두 도시에 대한 여행의 기록이다.

이 여행에서 다시 돌아온 자의 도시는 떠날 때의 도시가 아닌 다른 것이 되어 있을 것이다.

You are invited to

understanding culture
and negotiating boundary

"be my city"

ARCHITECTURAL EXCHANGE

한국과 스페인의 도시읽기

Date of Printing (발행일) 25 May, 2012 (2012년 5월 25일)

Published by (발행처) SPACETIME (시공문화사)
Publisher (발행인) Kim, Ki-hyun (김기현)
Date of Registration (등록) 12 March, 1993 (1993년 3월 12일)
Editing (편집) Yi, Soon-hak (이순학)

Address (주소) 5F Kukdong bldg., 200 Hyunjeo-dong, Seodaemun-gu, Seoul, Korea
(서울시 서대문구 현저동 200 극동빌딩 5층)
Tel. 82-2-3147-1212, 2323 Fax 82-2-3147-2626
http://www.spacetime.co.kr
spacetime@korea.com
ISBN 978-89-5592-230-2

Printed in Korea
12,000 KRW

FORWARD

Jose María Torres Nadal + Enrique Nieto

South Korea-Spain, Alicante- Seoul: the story of the work that is presented here starts in the beginning of 2010, when Jose María Torres Nadal is invited by Luis Fonseca, cultural aggregate of the Spanish Embassy by that time, to give a lecture in Seoul. That very moment Jose María meets Jorge Almazán and some other Spanish and Korean architects as Santiago Porras, Marc Brossa or Sanki-Choe, Assistant Professor of the Seoul University. They all wanted somehow to set cultural linkages. After this, at the end of 2010, and by means of Luis Fonseca, the 'Spain-Korea Symposium for Architecture and Urban Design: REPRESENTATIONS, SPACES AND PRACTICES FOR THE CULTURAL CONSTRUCTION OF AN ARCHITECTURE PEDAGOGY' is organized by Jose Maria Torres Nadal, Enrique Nieto, Marc Brossa, Santiago Porras, Jorge Almazán and Jungin Kim from the Soongsil University where the symposium takes place. After that the interest still remains and it is in 2011 when the workshop leading to this publication is configured.

It is, then, about a multiform experience that strives to set in time several models of relationship between two cultures and moments as different as the Korean and Spanish cases. Yet in order to make this kind of experiences happen it is necessary the existence of people with wider expectations acting as agglutinating agents and activating a certain curiosity. This is the case of Jungin Kim, whose presence and continuous interest have been fundamental.

The symposium brought to light our interest for enabling a conversation that claimed for the cultural role of the architectural project. Yet at this symposium we could see some relevant points for a later development of this relationship: the interest of the University of Alicante was and it is to open a discussion that enables the construction of a series of teaching events which are not ruled by the usual practices of teaching and learning that has been historically established as "architectural knowledge". At this symposium we encouraged the students to intervene and tell their impressions acting as a kind of alternative texts to the official stories of the "speakers". And we do think this little action and its related ideology provided both teachers and students with an "empathy" towards new methods that specially value a possible condition of friendship for running a workshop between so different cultures and individuals.

So far, the aperture of Europe to Asia has happened mainly within the field of the extensive-productive. We Spanish architects have conceived the Asian continent as a dignified solution to our crisis and our productive anxiousness, running away from an exhausted context as ours. Until now, this experiment has worked half way. Professional models do not seem to be easily exportable. They are usually determined by a net of personal relations that barely can be reduced exclusively to a matter of production. The contributions to this symposium tried to open this exchange up to other complexities, designing what in the coming future could be new areas of centrality within the intercontinental debate.

The whole trip experience of our students to Seoul, and vice-versa, was based on this friendly approach between the city, the citizens and our students themselves converted into a sort of Sherlock Holmes that investigate and try to find out the plot, the murderer and crime scene. The stay of each one at each other's homes goes beyond an economical reason. It was a way of saying that knowledge could only be established from the violence and contact, from the different forms of affections that take place within the everyday. Our idea is that this structure of events avoids the histrionic idea, relatedusually to architecture workshops, that students from somewhere fix the problems of somewhere else: we propose a situation closer to the work of artists working for a network of friends in order to build a net of confidence.

That idea of searching and enabling the controversy has been relevant. It has made possible the understanding of this dematerialized way of connection -that appreciates the accidental, the dispensable, the experiential- without previous codes or pre-conceived ideologies. It is something based on a universal language of the technology of networks. It is also something accepted by teachers and students as a starting point, as a way for building and as a way for creating strategies leading to an architecture whose aim is creating forms of happiness.

It is true this happens in a place whose backdrop is the stage of the great ideologies and macro decisions and political formations of the city. And it is also true that, within this macro reality, these actions are easily conceived as meaningless. Nevertheless, as time goes on, there are more arguments that are built from persistence, survival and transformation. Survival, because in architecture schools it is impossible to stay alien to this "ironic ethics" that, as the Marx's ghost, travels around the world; and transformation, because while we are not sure whether they want to talk to us, we are sure we want to talk to them as well as we know how and what we want to talk about.

Beyond, there is the future and the discovering of the sort of certainties these initial experiences have built; experiences where verbalization, teaching and production of project's culture meet a fragile and instituting context for new practices, and assuming a higher risk that official institutions barely support.

There it is the extreme importance of the role assumed by the involved universities and, specially, that of the Spanish Embassy in Seoul for making this series of events successful.

FORWARD

Jose María Torres Nadal + Enrique Nieto

Corea del Sur-España, Alicante-Seúl: La historia del trabajo que aquí se presenta, comienza a principios de 2010, cuando José María Torres es invitado por Luis Fonseca, por entonces agregado cultural de la Embajada española, a dar una conferencia en Seúl. En ese momento conoce a Jorge Almazán y a otra serie de arquitectos españoles y coreanos, como Santiago Porras, Marc Brossa o Sanki-Choe, Profesor Asistente en la *Seoul University*, todos ellos con ganas de estrechar vínculos culturales. Posteriormente, a finales de 2010, y por la mediación del mismo Fonseca, se organiza el *Spain-Korea Symposium for Architecture and Urban Design: REPRESENTATIONS, SPACES AND PRACTICES FOR THE CULTURAL CONSTRUCTION OF AN ARCHITECTURE PEDAGOGY*, que cuenta con la organización de Jose María Torres, Enrique Nieto, Marc Brossa, Santiago Porras, Jorge Almazán, y Jungin Kim, de la Soongsil University, lugar de celebración del simposio. A partir de ahí el interés continúa y es en 2011 cuando se configura el taller que da origen a esta publicación, dirigido por Iván Capdevila y Jungin Kim, y con la presencia de estudiantes coreanos y españoles.

Se trata, por tanto, de una experiencia multiforme que intenta fijar en el tiempo distintos modelos de relación entre dos culturas y dos momentos tan diferentes como son el coreano y el español. Pero para que este tipo de experiencias fructifique es imprescindible la presencia de determinadas personas que actúen como aglutinadores y activadores de una curiosidad y expectativas más amplias. En este caso la presencia e interés constante de Jungin Kim, de Soongsil University, ha sido fundamental.

El simposio puso de manifiesto nuestro interés por establecer una conversación que reivindicara el papel cultural del proyecto de arquitectura. Pero ya en este simposio pudo verse algún dato que ha sido relevante en el desarrollo posterior de la relación: que el interés de la Universidad de Alicante era y es abrir una discusión que permitaconstruir una estructura de sucesos docentes que no estén regulados por las practicas habituales de enseñar y aprender aquello que ha sido establecido como "conocimiento arquitectónico". En el simposio dimos pie a que los estudiantes intervinieran y explicaran sus impresiones como textos alternativos a las narraciones oficiales de los "speakers". Y entendemos que esa pequeña acción y la ideología que comporta dio forma a una "simpatía" de profesores y alumnos hacia métodos que valoran por encima de todo la condición de la amistad posible para montar un taller entre culturas individuos tan diferentes.

Hasta el momento, la apertura de Europa a Asia se ha producido mayoritariamente en el ámbito de lo productivo extensivo. Los arquitectos españoles hemos creído ver en el continente asiático una salida digna a la crisis que continuara nuestras ansias productivas, huyendo de un contexto agotado como el nuestro. Hasta la fecha el experimento ha funcionado sólo a medias. Los modelos profesionales no parecen ser exportables con facilidad. Llevan asociados una vastísima red de relaciones que no se dejan reducir con facilidad a cuestiones exclusivamente productivas. Las aportaciones al simposio intentaban abrir la experiencia de intercambio a otras complejidades, diseñando lo que en el futuro podrían ser nuevas áreas de centralidad en el debate intercontinental.

Toda la experiencia del viaje de los estudiantes coreanos a Alicante, y la correspondiente visita de nuestros estudiantes Seúl, se basó en esta aproximación amistosa, cordial, entre la ciudad, los ciudadanos,y los estudiantes convertidos en Sherlock Holmes que investigan, que tratan de averiguar el móvil, el asesino y el escenario del crimen. La estancia de cada uno de ellos en la casa de su compañero de trabajo era más que un tema económico. Fue una manera de decir que ese conocimiento sólo es posible establecerse desde la violencia y el rozamiento, desde las distintas formas de afecto que sólo sucedenen la vida diaria. Nuestra idea es que esta estructura de sucesos abandona el papel y la idea de aquellos talleres de arquitectura basados en la histriónica ficción de pensar que los estudiantes de un lugar resuelven algo de otro lugar: proponemos un trabajo más próximo al de un trabajo de unos artistas que trabajan para una red de amigos, para construir una red de confianza.

Esa idea de ir a buscar, de propiciar la aparición de la controversia, de pensar que siempre estamos nosotros y ella ahí, alcanzando un estado de deseabilidad, ha sido muy relevante. Porque ha permitido que esta forma desmaterializada de conexión, que valora lo accidental, lo prescindible, lo experiencial, fuera entendida sin códigos previos, sin ideologías interpuestas. Es algo que se basa en un lenguaje universal de la tecnología de las redes, algo que es aceptado por profesores y alumnos, como punto indiscutible de partida, como un modo de hacer obras, de crear estrategias que convocan a la arquitectura en la construcción de formas de felicidad.

Es cierto que esto ocurre en un lugar cuyo telón de fondo es el escenario de las grandes ideologías y las macro decisiones y formalizaciones políticas de la ciudad. Y es cierto que en esta macro-realidad estas acciones son fácilmente calificables como insignificantes. Pero cada vez son más los argumentos que se construyen desde la persistencia, la supervivencia y la transformación. La supervivencia porque para las escuelas de arquitectura es imposible permanecer ajenas a esta "ética irónica" que como el fantasma de Marx recorre el mundo; y a la transformación, porque si bien no estamos seguros de que ellos quieran hablar con nosotros, nosotros sí queremos y sabemos de qué y cómo hablar con ellos. Y luego queda el futuro, y descubrir el tipo de consistencias que estas experiencias iniciáticas han construido. Unas experiencias donde la verbalización, la enseñanza, y la producción de la cultura del proyecto encuentran un acomodo frágil e instituyente de nuevas prácticas, asumiendo un riesgo que las instituciones oficiales rara vez se encuentran dispuestas a apoyar. De ahí la radical importancia del papel de las universidades implicadas; y en este caso el de la Embajada de España en Seúl, para el éxito de esta sucesión de eventos.

대한 민국과 스페인, 알리칸테 서울:

이 이야기는 호세 마리아 토레스 나달교수가 주한 스페인대사관의 루이스 폰세카 문화영사의 주선으로 서울에서 강연초청을 받은 2010년으로 거슬러 올라간다. 당시 호세마리아는 호케 알마잔, 산티아고 포라스, 마크 브로사와 서울시립대의 최상기 교수를 만나게 되고 이 그룹은 어떤 식으로든 문화적 연결 망을 만들어 보고자 하였다. 이후, 2010년 말에 다시 루이스 폰세카의 주선으로 호세마리아, 엔리케 니토, 마크 브로사, 산티아고 포라스 그리고 숭실대학교의 김정인 교수가 모여 '건축과 도시를 생각하는 스페인 - 한국 국제 심포지엄: 건축교육의 문화적 형성과 그 표상, 공간, 그리고 작업'이라는 주제를 가지고 유럽과 한국의 학자들을 초청하여 심포지엄을 열게 된다. 이 대회 이후 논의된 주제에 대한 관심은 계속적으로 지속되었고, 2011년에 스페인 - 코리아 교류를 통해 이러한 출판이 가능하게 된 계기가 되었다.

이것은 한국과 스페인 간의 문화와 그 차이에 대한 몇 가지 건축적 모델을 가능케 하는 다양한 (건축적) 경험에 관한 이야기로 함축된다. 하지만 이러한 경험이 가능하도록 하기 위해선 폭 넓은 비전으로 제기된 주제에 대한 흥미를 지속적으로 끌어들이고 맞추어나가는 역할자의 존재가 필수적이었다. 이 경우에 가장 적합한 사람이 김정인 교수였고 그의 참여와 지속적인 관심은 이러한 교류에 있어 필수 불가결한 요소가 되었다.

심포지엄에서 우리의 관심이 표현된 주된 대화 내용은 건축적 작업의 문화적 역할이었다. 또한 이 심포지엄에서 우리는 후에 이루어진 이러한 교류를 위한 연관된 요소들을 발견하게 되는데 여기서 알리칸테 대학의 관심은 우리가 역사적으로 만들어져 온 "(특정한) 건축적 지식"을 학습하는 일반적인 상황에 지배 받지 않는 일련의 새로운 교육적 사건을 위한 열린 대화였다. 이 심포지엄에서 우리는 건축을 배우는 학생들이 정형화된 이야기의 전달자가 아니라 대안적인 텍스트로서 자신들의 개인적 인상과 이를 적극적인 참여형태로 보여줄 수 있는 행동을 격려하였다. 그리고 우리는 이러한 자그마한 시도와 이에 연관된 이데올로기가 교수와 학생에게 매우 상이한 문화와 그에 속한 개인 사이에서 건축적 교류를 가능하게 하는 호혜적 상황과 이를 위한 새로운 방법으로의 공감을 얻어낼 수 있었다고 생각한다.

지금까지 유럽과 아시아의 연결통로는 주로 확장된 (자본의) 생산 영역에서 주로 이루어져 왔다. 우리 스페인의 건축가들은 아시아 대륙을 우리처럼 지쳐있고 소진된 맥락을 피하기 위한, 즉 유럽의 위기와 (유럽적 자본의) 생산적 기우에 대한 훌륭한 해결의 땅으로서 생각해 왔다. 지금까지 이러한 시도는 그 기대에 당초 미치지 못했다. (건축적) 프로패션 모델 또한 쉽게 새로운 땅에 이식할 수 있는 것처럼 보이지 않는다. 건축적 작업은 보통 생산의 관계로만 축소 할 수 없는 사람들 사이의 관계망에 의해 결정된다. 이처럼 미래의 대륙간 교류의 장에서 이러한 교류의 경험이 여러 복잡한 (관계적) 상황에 열려있고 새로운 중심적 영역을 만들어 갈 수 있도록 기여케 하는 것이다.

학생들의 서울여행은 한국 학생들의 스페인 경험과 마찬가지로 도시와 도시민 사이의 호의적인 접근과, 학생들 자신이 셜록 홈즈와 같이 준비된 플롯을 발견하고 범죄의 현장을 조사하는 것과 같은 경험을 바탕으로 구성되었다. 상호 일대일로 파트너의 가정에서 숙박한 것은 단순히 경제적인 이유만을 고려한 것이 아니었다. 이러한 구성은 (건축적) 지식이라고 하는 것이 일상생활에서 일어나는 모든 것들, 어색함(Violence), 상호 접촉, 그리고 새로운 형태의 정서적 표현에서도 형성될 수 있다는 것을 대변하고 있기도 하다. 우리는 이러한 이벤트의 구성이 기존 형태의 워크숍, 즉 어떤 지역의 학생들이 주어진 어떤 지역의 문제를 풀어가는 전통적인 생각을 거부하는 형태라고 생각한다: 대신에 우리는 어떤 예술가가 신뢰를 구축하기 위해 친구들의 네트워크를 그리는 작업과 유사한 상황을 이 교류를 통해 제안한다.

논란거리를 가능하게 하고 발견하려는 생각은 (이러한 교류에) 연관되어 있다. 이것은 우연적이고 순간적이며 경험적인 것을 이해함에 있어서 이전의 (사회적) 코드나 이미 고안된 이데올로기의 제약없이 물질화되기 어려운 연결의 방식을 이해 가능하게 한다. 이것은 (정보의) 연결망 기술을 이용한 전세계적 언어에 기초한 어떤 것이며, 이것은 또한 교수와 학생들에 의해, 행복의 형태 구현을 궁극적 목표로 하는 건축의 전략을 만들어가는 출발점으로 받아들일 수 있는 그 어떤 것이다.

이러한 건축이 장대한 이데올로기와 거시적 결정 그리고 도시의 정치적 상황과 같은 배경을 피해 갈 수 없다는 것은 사실일 지도 모른다. 그리고 이러한 거대 현실 속에서 우리가 시도한 이런 교류와 건축적 실험은 의미 없어 보일지도 모른다. 그럼에도 불구하고 우리의 이러한 시도가 지속적으로 생존하고 변화해가는 모습에 의해서 더욱 더 많은 건축적 논제의 모습으로 시간이 흘러 갈수로 드러날 것이다. 생존이라고 말한 이유는 건축대학에서, 막스의 유령이 세계를 유랑하듯이 현재의 "아이러니한 윤리"에 무관하게 교육을 지속하는 것이 거의 불가능 하기 때문이고, 변화라고 한 이유는 우리가 (거대 현실속에 있는) 그들이 우리에게 말을 걸어 올 지 불분명하지만 우리는 그들에게 무엇을, 어떻게 말할지 알고 분명히 말을 걸고 싶어하기 때문이다.

이 모든 것을 넘어서, 이러한 초창기 경험이 만들어 내는 어떤 분명한 상황 발견과 그 (교육적) 미래가 있다. 그리고 이러한 경험은 (건축의) 작업문화를 생산하고, 표현하고, 가르치는 행위가 기존의 고착화된 교육제도가 포용하기 어려운 새로운 건축 작업을 위한 유연하지만 (현재와 미래의) 맥락에 적합한 (교육의) 상황을 만들어 가게 될 것이다.

이러한 시도에서 연관된 대학들이 각자에게 분담된 역할을 충분히 수행하였고 특별히 서울의 스페인대사관의 후원이 이 모든 연속적인 과정을 성공적으로 이루어내는 데 매우 중요하였다.

THE SITUATION

Ivan Capdevila

"Cities should generate, reflect, and activate life, their environment organized to precipitate life and movement. Situation, the happenings within spaces in the city, the transient throw-away objects, the passing presence of cars and people are as important as- possibly more important than - the built demarcation of space. Situation can be caused by a single individual, by groups or a crowd, their particular purpose, occupation, movement or direction. Situation can be traffic, its speed, direction, classification."[1]

More than half century has passed by since these words by Constant, nearly paraphrased by Archigram, were claiming a new way of understanding the architectural fact, more in a phenomenological sense than in a tectonic one.

These words did not happen in an isolated context as they were preceded by several years of derive[2] along the streets of Paris, psycogeography, sociological texts and, in short, by a search for creating "ambients of situation"[3].

All these approaches showed a need for a change of architectural paradigm[4]. The ultimate aim of the new architecture, based on assuming reality "as found"[5], was to generate ordinary situations[6] in which the user, individual or collectively, participated passive or actively.

However, the architectural learning kept on being ruled by criteria far away from a direct experience of reality itself.

It was not until 1977, with Learning from Las Vegas by Robert Venturi and Dennis Scott Brown and, specially until 1978, with the publication of Delirious New York by Rem Koolhaas, that we could understand that reformulation of reality had to be made by previously learning from its direct experience and, what is more, from its detailed restitution.

Rem Koolhaas would formulate his architectural thinking through a detailed observation and description about how Manhattan acquired its urban condition. And, what is more important, he helped us to understand that the urban can be explained not only from the quantitative and big events but, and especially, from the qualitative and small events.

Since then, there have been many studies that have taken the phenomenology of urban micro events as the place for urban knowledge. We could highlight from the example of Made in Tokyo by Atelier Bow-Bow to the recent case studies by Zira02 about "corralas"[7] in Madrid.

This ideology is known, shared[8] and, in many occasions, has been put into practice by the AeA[9].

The workshop is conceived, then, in continuity with this tradition and, what is more, taking into account the particularity of the experience itself. Such far away placeslike Alicante and Seoul, not only physically but culturally and emotionally, turn this fact into an opportunity to lead the workshop to a cultural mutual learning based on a direct experience of a strange and unknown reality.

In other words, drawing images of buildings is a common and repeated situation. However, learning architecture from a direct experience of situations, in which it is manifested, is an unusual situation and, in this particular case, unique. This is the reason why the workshop had to be celebrated on the street[10] and not in the classroom. Situations are out there and our work is to live them in order to, after that, learn from them.

Besides, we are convinced that the mere fact of detecting the ordinary turns it into extraordinary. And it is the foreigner who has a bigger ability of scrutinizing reality, as he can more easily distinguish which aspects of the ordinary are extraordinary.

Thus, accepting the fact that architecture is a vital experience[11] and that our ultimate goal is to propose it, our first goal should be to live and experience it.

Following this thesis, the only condition for the selection of urban situations - case studies is that "the ordinary is collectively celebrated".

On the other side, our methodology will be that of moving around the city under a permanent state of derive[12] and flow[13], as well as with a totally "play"[14] attitude.

The work will be not only that of living and experiencing such situations. Students will have to understand how these situations are built, in the sense of finding out the relations between culture, reality, space and situation.

We do believe architectural learning is not only produced indirectly through others - books, readings, etc.- in a more or less rational way, but direct, emotional and abruptly.

When making an effort to face the unknown, trying to understand it, draw it, measure it, etc., it is the very moment when learning happens.

The put into practice of this learning can be immediate or quite the opposite. This is not relevant. The most important aspect is that it will be something that any student will have incorporated to his personal baggage. And yet more important, instead of projecting more or less 'a certain architecture', he will have had the chance of, at least, really living 'other architecture'.

In other words, instead of the transfer of a method, the workshop is focused on the transfer of an attitude.

This is our aim.

---

1 Situation is a concept that finds its origin in Jean-Paul Sartre.

2 Situationist concept, the derive is a way of using the urban space and rediscovering its qualities, letting it go, wasting time, etc.

3 The thinking of the 'InternationaleSituationniste' brings the two fundamental problems happening within the city in the 60's: the position of the individual in the anonymous modern city and the exponential increase of mass media that led Debord to write "The Society of Spectacle" as a new condition. On the most basic level, the resistance against the modern funtionalist city was structured according to "human relations". The funtionalist city was seen as too mechanical and anonymous to incorporate such things as children playing in fountains (Robert Venturi), changing conditions of life (Alison and Peter Smithson), everyday informal social relations (Jane Jacobs) and the passion of human life (Guy Debord). Human being is the centre. It is not any more an expression of modern and rational life, but an individual entity willing human contact, a sense of place and, especially in the situationist thinking, adventure.

4 Following the most contemporary definition, by Thomas Kuhn, we understand paradigm as the group of practices that define a discipline during a period of time.

5 Terminology coined by Allison y Peter Smithson to refer to the intrinsic qualities of reality as it was, with all its virtues but also with all its defects.

6 The very moment Allison and Peter Smithson presented reality itself, trimming and sticking people figures at the front of their photomontages for the proposal for the Golden Lane competition in 1952, there was a pile of

상황

이반 카프데비아

도시들은 삶을 일으키고 반영하며 활성화시켜야 하며 이 도시들의 환경은 유동적인 삶을 촉진시키도록 구성되어진다. 상황, 즉 도시 내 공간에서 일어나는 사건(일시적으로 낭비되는 대상, 지나쳐가는 차와 사람들)은 주어진 경계의 공간 만큼 중요할 수 있다. 상황들은 개인, 단체 혹은 군중의 특정한 목적, 직업, 움직임과 방향에 의해 야기될 수도 있고 교통, 교통의 속도, 방향, 분류일 수도 있을 것이다.[1]

아키그램에 의해 재구성된 콘스탄의 언어들이 건축의 구축적인 의미보다는 현상학적 의미로서 건축적 사실을 이해하는 새로운 길을 연 이후 어언 50년 이상의 세월이 흘렀다.

여기에 쓰인 말들은 파리의 거리를 따라 간 몇 년의 유랑[2]을 통해서, 심상지리적이며 사회학적 글이다. 한마디로, "상황적 분위기[3]"를 만들기 위한 선행된 리서치로서의 결과이지 고립된 정황에서 발생된 것이 아니다. 이 모든 접근방법들은 건축적 패러다임[4]의 변화를 위한 필요성을 보여주는데, 즉 '이미 성립된 것[5]'으로 가정된 새로운 건축의 궁극적인 목적은 수동적이거나 적극적으로 참여한 개인 혹은 다수의 사용자가 참여한 평범한 상황[6]을 만들어내는 것이었다.

하지만 건축적 배움은 현실세계의 직접적 경험으로부터 멀어진 규범들에 의해 계속적으로 지배받아왔다. 1977년의 로버트 벤츄리와 데니스 스코트 브라운의 "Learning from Las Vegas", 특히 1978년의 렘 쿨하스의 "Delirious New York"의 출판으로 인해 우리는 현실의 재형성이 직접적 경험과 그것의 상세한 재구성에 의해 이전에 학습된 것에 의해 만들어져야 한다는 것을 알게 되었다. 렘쿨하스는 어떻게 멘하튼이 도시의 컨디션을 획득했는지 상세한 관찰과 서술을 통해 그의 건축적 사고를 공식화 했다. 그리고 이에 더해 더 중요한 것은 그는 도시성이 정량적이고, 큰 이벤트들 뿐만 아니라 정성적이고 작은 이벤트에 의해 설명될 수 있다는 것을 이해하도록 도와주었다. 그 이후로, 도시에 대한 지식을 얻기 위한 장소로서 도시의 미세한 사건에 대해 현상학적 견해를 가진 수많은 연구가 있었다. 실례로서 우리는 아틀리에 바우와우의 "Made in Tokyo"에서부터 Zira0에 의해 연구된 마드리드의 "corralas"[7]에 대한 최근의 연구를 조명할 수 있을 것이다.

이 이데올로기는 잘 알려져 있고 공유되었으며[8] 또한 많은 경우에 AeA의 의해 작업화 되었다.[9] 이러한 면에서 이 워크샵은 이러한 건축적 전통과 경험이 가지는 특수성을 고려하는 것으로 착안되었다. 알리칸테와 서울처럼 물리적으로, 또한 문화적, 감성적으로 멀리 떨어진 장소는 우리의 워크샵을 통해 서로의 색다르고 잘 알려지지 않은 서로간의 상황을 직접적으로 경험함으로서 상호 문화에 대한 배움의 기회로 이끌게 될 것이다.

바꾸어 말하면, 빌딩 이미지를 그리는 것은 일반적이고 반복된 상황이다. 하지만 드러난 상황들의 직접적인 경험을 통해 건축을 배운다는 것은 일반적이지 않은 (교육적) 상황이며 우리의 경우에 특히 유일하다. 이는 워크샵이 교실이 아닌 도시의 거리[10]에서 행해져야만 하는 이유이다. (건축적)상황들은 도시곳곳인 외부에 있으며 우리의 작업은 그러한 상황들로부터 건축적 배움을 얻기 위해 그곳에 더불어 사는 것이다. 게다가, 우리는 단지 평범함을 찾는 일에 불과한 것처럼 보이는 것에서 어떤 특별함을 발견할 수 있는 가능성을 확신하게 되었다. 그리고 그것은 현실을 면밀히 새롭게 보는 더 큰 능력을 보유하게 된 외국인 (이방인) 에 의해서이다. 왜냐하면 그들은 내국인이 평범하다고 느끼는 것에 특별한 측면이 있다는 것을 (외국인이기 때문에) 쉽게 구별 해 낼 수 있다.

이와 같이, 건축은 살아있는 경험이다[11]라는 사실을 받아들이는 것 그리고 우리의 궁극적인 목표는 이것을 제안하는 것이며, 따라서 우리의 첫 번째 목표는 그곳에서 생활하고 경험해 보는 것이다.

이 논제를 따라, 도시적 상황의 선택을 위한 유일한 조건은–각 장소의 연구가 '평범함을 다같이 행하고 기릴 수 있다'를 만족하는 것이다.

반면에 우리의 방법론은 유랑[12]과 흐름[13]의 지속적 상태에서 완벽하게 즐기는 태도(노는 태도)[14]를 가지고 도시를 여기저기 돌아다니는 것이 될 것이다.

우리의 작업은 도시적 상황을 경험하고 살아보는 방법적인 측면만은 아닐 것이다. 학생들은 문화, 현실세계, 공간과 상황 사이의 관계를 알아내는 방식으로 이러한 상황들이 어떻게 이루어지게 되었는지 이해해야만 할 것이다.

우리는 건축적 학습이 책과 글 등을 통한 간접적인 방법뿐만 아니라 다소 덜 합리적인 방법처럼 보일지라도 직접적이고 감정적이며 뜻밖의 경험으로 생산되어질 수 있다는 것을 확신한다.

알려지지 않은 것과 마주치고, 그것을 이해하려고 노력하고, 그려내고, 측정하는 등의 노력을 할 때 그 순간이 바로 건축적 배움이 일어나는 시간이다. 이 배움을 실행에 옮기는 것은 즉각적으로 일어나거나 아니면 이와 상반된 것이다. 이는 우리의 작업과 별 연관을 가지지 못한다. 가장 중요한 측면은 그것이 어떤 학생들에게는 그의 개인적 짐으로 묶여져야만 하는 무언가가 되어 있을 것이다. 그리고 더 중요한 것은, 다소 일반적이고 분명한 건축을 그리는 대신에 최소한 그는 실제로 '다른 건축'에 거주하는 기회를 가질 거라는 것이다.

다시 말해, 방법의 변화 대신에 이 워크샵은 (건축에 대한) 태도의 변화에 초점을 맞추고 있다. 이것이 우리의 목표다.

---

architectural proposal happening in the following decades. In all of them, the everyday reality, "as found", negative, exagerated, scaled or no matter how it happened, was the aim to reach.

7 The "corrala" is a typical architectural typology from traditional Madrid. It is used to describe a housing only facing an inner courtyard and located in the interior of a urban block.

8 The main thesis of the courses I ran in 2009-10, along with Andres Jaque, and in 2010-11 supported the idea that projective ability is directly related with descriptive ability and restitution of reality.

9 School of Architecture.University of Alicante.

10 The reference to the street is taken as a figuration of those places where the ordinary is celebrated collectively.

11 As Cristiano Toraldo di Francia says: "the intellectual does not guide the user nor the contrary. He becomes the user instead". See: Ver: MuseoValenciano de la Ilustracion y la Modernidad.,Arquitectura radical : MUVIM, from 7th November to 2nd December 2001. [Valencia]: GeneralitatValenciana, 2001.Pp 205-222

12 In a Situationist sense.

13 Mental state in tune with the emotional and urban fluxes where the individual enters, being leaded by a non controlled attraction for the unknown, the unexpected, the undetermined, the unplanned, for what is happening next; sometimes, it is not correct. This is, nevertheless, part of the adventure the Situationists defended as a methodology.

14 Johan Huizinga defines play as a universal human impulse that is essentially the basis of all culture. See: Johan Huizinga, Homo ludens (Madrid;Buenos Aires: Alianza Editorial ;Emecé Editores, 1972).

CONTENTS

UNDERSTANDING CULTURES

Jungin Kim

This Korea-Spain exchange as its subtitle proclaimed has been a trial ground for thinking and learning design by negotiating our attitude towards "culturally different" urban places. As Don Mitchell suggested, the word of "culture" brings incredibly complex connotations.[1] Thus, this wide and ambiguous term for our specific purpose necessitates some clarification. In the rapidly rising information society all over the urban centers of the globe, talking about "culture" under the current mode of speed and communication may sound obsolete, let alone toned in anachronism or at best in nostalgic anti-universalism.[2] On the other hand, there is yet another rising demand to emphasis regional differences and local identities against 'homogenizing' forces of current globalization. From the street protest by a small shop owner to the militant nationalist movement, we come across everyday news of numerous local struggles that are never in support for the overarching spaces of globalization. Useful for our intention under this binary contradiction is to locate the word, "culture" among the discursive frameworks of multi-variant global urbanism, and puts forward our position about this term."[3]

That is, we like to think about "culture" as a dialectical condition between the universal or the global space that has undergone historically exceptional moments in a given urban place, and at the same time the individual space of lived urban experience that constantly adopts, negotiates, and appropriates the larger episteme of the time, the current phase of informational age. "Culture" from this particular reading is thus defined both as 'produced' and 'productive force' that binds our lived experiences in the city that is more and more interconnected globally.

Relating this dialectical construction of "culture" to our thinking of urban spaces, this exchange between Spanish and Korean students becomes a trial ground and an exploratory exercise to recognize each other's particular view of the city as a "cultural" construction. In so doing, the students from Spain and Korea had not only visited each other's urban place but also worked in the virtual space, producing common web pages and interacting through social networking media. This global environment is also an opportunity for those participated to widen the current view of architectural learning and thereby to rethink the "conventional" studio curricular operative in many design institutions. As "culture" being the critical tool of engagement with urban space, the entire exchange has worked as a platform that influences and is influenced by our actions- in our case, experiencing and designing action in each other's city - making each other's "invisible" urban space, or rather urban culture, "visible".

By postulating relationship between "culture" and urban spaces this way, the Spain Korea workshop becomes trailblazing exercises in order to find out, despite the geographical distance and scalar difference between Seoul and Alicante, a resourceful ground that awaits new spatiotemporal potentials in the imagination of (future) architecture and urbanism. Through this opening, we hope that universally predictable mode of "the urban" could reveal its hidden dimensions, or liminal conditions that contain unpredictable, but amicable encounters of the "cultures" between Spain and Korea.[4]

1 Don Mitchell categorizes various reading of culture in two basic ways: one as a "total way of life" of people, encompassing the dwellings, housing styles, private and public space of the city and most importantly *values.* On the other hand, the term indicate certain things: works of art, musical production, the stuff that gets put in museums and concert halls, or popular version of things that occur in clubs and stadiums, that appear in the commercial strips. Don Mitchell, *Cultural Geography* (Oxford, Blackwell publishing, 2000), p14.

2 See the comparison between "civilization" and "culture" made by Alan Colquhoun in G.B. Nalbantoglu and C. T. eds. *postcolonial space* (New York, Princeton Architectural Press, 1997), pp 13-23.

3 At this juncture, I agree with Don Mitchell's arguments when hesays, "while there is no such thing as culture, our fights over culture are deadly important." Don Mitchell,Ibld. p12.

4 As cultural anthropologist, ÁniwaOng indicated this view of the city as a site of experimentation which allows us "to integrate qualities of fluidity, interactivity, and interactivity that crystallize the possibilities within which we reimagine, remake, and re-experience urban conditions and the notion of the urban itself". See Ananya Roy and AhiwaOng eds.*Worlding Cities: Asian Experiments and the Art of Being Global*(West Sussex, Wiley-Blackwell, 2011),p10.

NEGOTIATING BOUNDARIES

Jungin Kim

"The skyline rises in the East" - Rem Koolhaas

Throughout the semester, Korean students in the Workshop challenged the limiting use of "culture", operated as semantic markers that set one people off from another while reminding us of our membership. Positioning ourselves in this position, students reexamined the current mode of urban knowledge and thought over what it meant to learn and practice architecture in Seoul, the Asian mega city. For the past decades, Seoul and its vicinity with its 24million people has been categorized often as a "megacity", the meaning of which only refers to its big size in the emerging hierarchy' of world cities. The city was distanced by a more glitzy and privileged term of "global city", which denoted the possession of power and influence generated from the urban centers, like New York or London. However, under the condition that "Asia becomes the metonymy for global capital", Seoul as an exemplar Asian city that preceded "Shenzhen Speed" takes the long debated Asiatic Mode of Production (and consumption)into the case of serious reevaluation. From the 1960's and spanning only a single generation, Seoul has demonstrated the structural transformation of urban society that are so familiar to us in the similar processes now happening in Chinese cities: the Asian megalopolis grown from unreasonable concentration of scarcity and under developments, to equally unreasonable global city with showcases of urban megalomania and hyper-developments.[5]

At the same time, we wanted to problematize the normalized spaces of geographical knowledge through our experiencing exercise. For us "Asia" as we know it becomes an inflexible and problematic concept as a conventionally categorized boundary in which a stereotyped appreciation of the "Asian culture" and the emphasis on its distinctiveness will miss the mark; instead, the city experience operative in this exchange was aiming at questioning "the privileged link between modernity and certain kind of (textbook) cities"[6]. The argument that "Asian" cities, and their skylines are rising gains stronger appeal when images of sprouting "instant cities" in the region seem to stretch the discussion of "the urban" as the emerging post national, post modern, and post colonial condition for density and diversity. This global condition creates the milieu for our Spain- Korea exchange that we are in high time to cast doubts on the current regime of architectural and urban knowledge and its colonizing power over our conscious. "Asia" as an idea, not merely has been a product of certain privileged knowledge for ordering the global spaces, but now can be a productive center that ruptures and transforms this knowledge.

The images of 'miraculous' contemporary cities and their high modernist urban discourses supply the long-familiar urbanism to us. They often shows surprisingly rationalized spaces that bear a close resemblance to the "panoptic representation(s) of the city", regularizing the everyday that was vibrantly lived.[7] While negotiating our inherited boundary of the traditional urbanism that is inscribed like a textbook, we feel that there are changing clouds and rising opportunities that our experience and interpretations - both in the cities of Alicante and Seoul - can create cracks to put some alternative readings of "the urban", and thereby open up the notions of "different space" that provincializes historically granted geographical declarations, Europe versus Asia, or particularly in this case, "global" cores versus "mega" peripheries. This is our call to return to 'other geography' and our way of critiquing the current production of urbanism both in Europe and Asia, which has often been obscured by the overly bloated spectacles of the metropolitan centers. In favor of "Urban Revolution" as coined by Henry Lefebvre, this urban experience and designing exercise starts from the reflexive positioning that urban knowledge learned from popular norms and hierarchies is ushered in only to reproduce and amplify the existing urban models equated in the "bureaucratic society of controlled consumption"[8].

Perhaps, one of the distinctive characteristics in the Korea - Spain exchange is an effort to show this uneasy awareness, and we as students of urban cultures want to test a new way of learning by going beyond the given boundaries of the existing urbanism, over and above the presumed urban narrations already set concrete by our institutional spaces.

5 Ibid, 321
6 Ibid, 309
7 Robin Vissar, *Cities Surround The Countryside: Urban Aesthetics in Postsocialist China*(Durham, Duke University Press Books, 2010) p28.
8 Henry Lefebvre, *Urban Revolution* (Minneapolis, University of Minnesota Press), p163.

ENTENDIENDO LAS CULTURAS Y NEGOCIANDO LOS LÍMITES:

Jungin Kim

Este intercambio Corea-España, como su subtítulo proclama, ha sido un espacio de prueba para el aprendizaje y la reflexión sobre el proyecto de arquitectura mediante la negociación de nuestra actitud respecto a espacios urbanos "culturalmente diferentes". Como Don Mitchell sugeriría, la palabra "cultura" genera connotaciones increíblemente complejas.[1] Así, este amplio y ambiguo término necesita ser clarificado en el contexto de nuestro propósito. En la creciente sociedad de la información que habita todos los centros urbanos del mundo, hablar de "cultura" bajo el modelo actual de velocidad y comunicación puede sonar obsoleto, incluso de manera aislada puede llevar al anacronismo o, en el mejor de los casos, a un anti-universalismo[2] nostálgico. Por otro lado, todavía hay un demanda creciente por enfatizar diferencias regionales e identidades locales en contra de las fuerzas "homogeneizadoras" de la globalización actual. Desde la protesta callejera de un pequeño tendero a un militante movimiento nacionalista, pasamos por noticias cada día de numerosas batallas locales rara vez a favor de los espacios generalizadores de la globalización. Bajo esta contradicción binaria, resulta útil en nuestro empeño situar la palabra "cultura" entre las estructuras discursivas de un multi-variante urbanismo global, y desarrollar nuestra posición respecto a este término[3].

Esto es, nos gustaría entender la "cultura" como una condición dialéctica entre el espacio universal o global que, históricamente, ha experimentado momentos excepcionales en determinados entornos urbanos y, al mismo tiempo, en espacios individuales de una experiencia urbana vivida que constantemente adopta, negocia y se apropia de la vasta episteme del momento: la fase actual de la era informacional. La "cultura" entendida desde esta perspectiva se define, así pues, como una "fuerza productiva" así como "producida" que unifica nuestras experiencias vividas en una ciudad cada vez más interconectada globalmente.

Relacionando esta construcción dialéctica de "cultura" con nuestro entendimiento de los espacios urbanos, este intercambio entre estudiantes españoles y coreanos se convierte en un lugar de prueba y un ejercicio experimental para reconocer la visión particular que cada uno tiene de la ciudad como construcción "cultural". Haciendo esto, los estudiantes de España o Corea no sólo han visitado sus respectivos espacios urbanos sino que también han trabajado en el espacio virtual, produciendo páginas web comunes e interactuando a través de las redes sociales. Este entorno global es también una oportunidad para los participantes de ampliar la visión actual del aprendizaje de arquitectura y, así, replantear el "convencional" sistema docente de muchas escuelas de arquitectura. Siendo la "cultura" la herramienta crítica de implicación con el espacio urbano, el intercambio en su conjunto ha funcionado como una plataforma que influencia y ha sido influida por nuestras acciones - en nuestro caso, diseñando y experimentando diversas acciones en las respectivas ciudades - haciendo "visibles" los espacios urbanos, o mejor dicho la cultura urbana, a priori "invisibles".

Postulando la relación entre "cultura" y espacios urbanos de esta manera, el taller España - Corea se convierte en un ejercicio pionero en el hallazgo, a pesar de la distancia geográfica y escalar entre Alicante y Seúl, de un espacio inventivo que aguarda nuevos potenciales espacio-temporales en la imaginación de una futura arquitectura y urbanismo. A través de esta apertura, esperamos que el modo universalmente predecible de "lo urbano" pueda revelar sus dimensiones ocultas o condiciones intermedias que alberguen impredecibles aunque amigables encuentros entre las "culturas" de España y Corea.[4]

1 Don Mitchell categoriza varias lecturas de la culturabásicamente de dos maneras: una, como un "modo total de vida" de la gente, abarcando viviendas, tipos de casas, espacio privado y público de la ciudad asi como la mayoria de aspectos relevantes. Por otro lado, el término indica ciertas cosas: trabajos de arte, producción musical, las cosas que se muestran en museos y auditorios o la versión popular de cosas que ocurren en clubs y estadios, que aparecen en zonas comerciales. Don Mitchell, Cultural Geography (Oxford, Blackwellpublishing, 2000), p14.

2 Ver la comparación entre "civilización" y "cultura" hechas por Alan Colquhoun en G.B. Nalbantoglu y C. T. eds. postcolonialspace (New York, Princeton ArchitecturalPress, 1997), pp 13-23.

3 En este punto, estoy de acuerdo con Don Mitchell cuando dice "mientras no existe tal cosacomo cultura, nuestros enfrentamientos acerca de la cultura son terriblemente importantes." Don Mitchell, Ibld. p12.

4 Como antropólogo cultural AniwaOng indicó esta visión de la ciudad como el lugar para la experimentación, la cual nos permite "integrar cualidades tal como fluidez, interactividad, e interactividad que cristaliza las posibilidades en las que re-imaginamos, re-hacemos y re-experimentamos condiciones urbanas y la noción propiamente urbana." VerAnanya Roy yAhiwaOng eds. WorldingCities: AsianExperiments and the Art of Being Global (West Sussex, Wiley-Blackwell, 2011), p10.

"La línea del horizonte se eleva en el Este" - Rem Koolhaas

A lo largo del semestre, los estudiantes coreanos desafiarían el uso limitado de la "cultura", la cual operaba a modo de indicadores semánticos y desvinculaba los unos de los otros mientras nos recordaba nuestra afiliación. Posicionándonos así, los estudiantes reexaminaron el modelo actual de conocimiento urbano reflexionando sobre qué significa aprender y ejercer la arquitectura en Seúl, la megapolis asiática. Durante las décadas anteriores, la metropolitana Seúl (con sus 24 millones de personas ha sido categorizada a menudo como una "megaciudad", el significado de la cual sólo se refiere a su gran tamaño en la emergente jerarquía de las ciudades del mundo. La ciudad se distanció gracias al término más deslumbrante y privilegiado de "ciudad global", evidenciando su poder e influencia sobre otros centros urbanos como Nueva York o Londres. Sin embargo, bajo la condición que "Asia se convierte en la metonimia para las capitales globales", Seúl, como un ejemplo de ciudad asiática que precede la "Velocidad Shenzhen", asume el ampliamente debatido modelo asiático de producción (y consumo) para su seria reevaluación. Desde los años sesenta y en una sola generación, Seúl ha demostrado la misma transformación estructural de la sociedad urbana que hoy nos son tan familiares en procesos de características similares en ciudades chinas: la megalópolis asiática crecida desde la escasez y subdesarrollo hasta una ciudad global igualmente irrazonable con muestras de megalomanía urbana y desarrollos desmesurados.[5]

Al mismo tiempo, queríamos problematizar los espacios normalizados del conocimiento geográfico a través de nuestro ejercicio experiencial. Para nosotros "Asia", tal como la conocemos, se convierte en un concepto inflexible y problemático en el sentido de un límite convencionalmente categorizado, en el que una apreciación estereotipada de la "cultura asiática" y el énfasis en su distinción nos llevará a perder su rastro; en su lugar, el operativo experiencial urbano de este intercambio lograría cuestionarse "el vínculo privilegiado entre modernidad y ciertos manuales de ciudades."[6] El argumento de que las ciudades "asiáticas" y sus perfiles están creciendo gana una mayor atracción cuando las imágenes de "ciudades instantáneas" brotando en la región parecen alargar la discusión sobre "lo urbano" como la emergente condición post-nacional, post-moderna y post-colonial para la densidad y la diversidad. Esta condición global crea el medio para nuestro intercambio España-Corea en el momento justo para proyectar dudas sobre el actual régimen de conocimiento urbano y arquitectónico y su poder colonizador sobre nuestra consciencia.

"Asia" como una idea, no sólo ha sido un producto de determinado conocimiento privilegiado para ordenar el espacio global sino que ahora puede ser un centro productivo que rompe y transforma su conocimiento.

Las imágenes de ciudades contemporáneas "maravillosas" y sus discursos urbanos altamente vinculados al Movimiento Moderno nos proveen de un urbanismo tan y tan familiar. A menudo nos muestran espacios sorprendentemente racionalizados que comportan una gran parecido con las "representaciones panópticas de la ciudad", regularizando la cotidianeidad que era vivida tan vibrantemente.[7] Mientras negociamos nuestro límite heredado del urbanismo tradicional de manual, sentimos la presencia de tiempos cambiantes y nuevas oportunidades para que nuestra propia experiencia e interpretaciones - tanto en el caso de Alicante como Seúl- pueda crear fisuras y añadir lecturas alternativas a "lo urbano" y, de este modo, dar paso a nociones de un 'espacio diferente' que provincializa declaraciones geográficas históricamente asumidas: Europa vs Asia o, 'particularmente' en este caso, núcleos "globales" vs "mega" periferias. Esta es nuestra llamada para volver a "otra geografía" y nuestra manera de criticar la producción actual de urbanismo tanto en Europa como en Asia, el cual ha sido a menudo oscurecido por los espectáculos excesivamente exagerados de los centros metropolitanos.

A favor de una "Revolución Urbana", como acuñaría Henry Lefebvre, esta experiencia urbana y proyectual comienza desde la postura crítica de que el conocimiento urbano aprendido de normas y jerarquías da paso a una mera reproducción y amplificación de modelos urbanos existentes equiparables con la "sociedad burocrática del consumo controlado."[8]

Quizá, una de las características más distintivas del intercambio Corea-España es un esfuerzo por mostrar esta difícil toma de conciencia y que nosotros, como estudiantes de culturas urbanas, queremos ensayar una nueva manera de aprender yendo más allá de los límites establecidos del urbanismo existente, sobre y por encima de las presuntas narraciones urbanas ya establecidas por nuestros espacios institucionales.

---

5 Ibid, 321
6 Ibid, 309
7 RobinVissar, CitiesSurroundTheCountryside: UrbanAesthetics in Postsocialist China (Durham, DukeUniversityPressBooks, 2010) p28.
8 Henry Lefebvre,UrbanRevolution (Minneapolis, University of Minnesota Press), p163.

LIVING THE URBAN LIFE

Alicante/by Ivan Capdevila

If we assume the mere fact of detecting the ordinary turns it into extraordinary, we will be able to understand why the workshop proposes to instrumentalize the trip itself.
According to this ideology, we believe the foreigner has a bigger ability of scrutinizing reality, as he can more easily distinguish which aspects of the ordinary are extraordinary.
This is, then, our work basis.
From here on, the criterion for selecting the sites-case studies is that "the ordinary must be celebrated collectively".
However, before understanding the collective ordinariness it is important to understand the domestic one. This is the reason why every local student must take in a foreign one.
Therefore, the foreign student aims to perceive some projections of the domestic space towards the public, as well as some characteristic cultural features that determine particular uses of the public space.
As a requirement by the Soongsil University, sites had to be at least 10,000m^2.
Our only condition for these sites was that they were understood in a broader sense beyond their physical contexts. This broader context should be that defined by their urban phenomenology.
According to these premises, the following brief is proposed:

1st Seoul - Alicante Workshop
BE MY CITY_ Exchanging the URBAN knowledge

A. IDEOLOGY:
Experiencing is intepreting. Interpreting is designing.
B. METHODOLOGY:
Several urban experiences will be designed. They will happen in a week time period. The students will learn from "what is already built". Guided by the locals, the strategy is to experience reality in person.
Passiveness is not allowed. It is not only about being in a certain place. The whole experience must be registered in detail by means of drawings, multimedia and so on...
C. MOTTO:
Culture for the people and vice-versa This motto's spirit is that of understanding culture from two opposite points of view.
According to the Oxford's dictionary definition, culture is both:
1) the arts and other manifestations of human intelectual achievement regarded collectively
2) the ideas, customs, and social behaviour of a particular people or society
The implicit aim behind the sites' choice is, precisely, to understand how both manifestations of culture influence each other. In this case, both sites are characterized by including within their boundaries that "programmed" and "recognized" culture but, at the same time, that other spontaneous culture: urbanely collective, socially Mediterranean and, sometimes, hidden and denied.
At the same time, both places are clearly different: one, is an existing place consolidated by history; the other, is an empty place, with no history nor identity, waiting for a better moment to be invented.
D. SITES:
SITE 1: "Las Cigarreras" cultural centre
This is a former cigarette factory inserted within the urban pattern that has been transformed into a cultural centre.
Exhibitions, concerts, movie projections, installations and so on provide this environment with a "programmed" cultural character.
However, less than hundred meters away, there is the Benacantil, Alicante's most emblematic mountain as on its top there is the "Santa Barbara" castle (9th century). At its north hill, directly facing "Las Cigarreras", thousands of youngsters climb up at night in the weekend to celebrate one of the most controversial "cultural manifestation" nowadays: the "Botellon". This consist of sharing conversations, memories, anecdotes amongst friends while having some spirit drinks. It is as reproducing any given situation inside a bar but outdoors.
The challenge would be, then, to understand how to experience personally both moments - walking at day time around the castle and standing up while holding a drink at night time- can determine a new way of comprehending the urban relationship between art and its places.
The designed experience to get to know this first site would consist of three acts:
a) Morning in the Castle
b) Evening at "Las Cigarreras"
c) Night in the "Botellon"

Therefore, can the sum of these three moments modify the way of designing a cultural centre? The students will manage for sure.

SITE 2: "La Goteta"
"La Goteta" is the place of a former urban quarry located close to the beach. Nowadays it is waiting for the construction of a "bright" Congress Hall.
The site itself, is an obligatory passing by point both for thousands of locals, living in the working class neighborhoods in the North, on their way to the "Postiguet" beach and for those living in the outskirts that reach the city centre by tramway.
Its coming future can bring lots of paradoxical situations as that of dressed up people attending the congresses getting mixed with those locals on their way to the beach wearing their swimming cloths and holding their umbrellas, towels and so on. This situation highlights its hybrid condition.
How can the fact of experiencing our Mediterranean beach affect the way of understanding a Congress Hall?
The designed experience would consist of four acts:
a) Return journey, passing by the site, to the Muchavista beach by tramway.
b) Hiking to the Serra Grossa mountain in the morning + picnic up there
c) Visit to the site
d) Evening-Night at the Postiguet beach

Therefore, can the sum of these four moments modify the way of designing a cultural centre? The students will manage for sure.
In short, all these events physically coexist although they seem to deny each other.

The subject's goal for this workshop lays on the basis that a future design (no matter what it is) forces these events to coexist as well as to learn from each other.
On the other hand, we should not forget that fun, joy, excitation and so on are not only suitable strategies for running a workshop but for self-learning and designing. That is very important, isn't it?

만약 우리가 단지 일상적 상황이 뭔가 특별한 상황으로 바뀐다고 가정한다면 왜 이 워크샵이 여행의 경험을 중요시하는 것을 제안하는지 이해할 수 있게 될 것이다.
우리의 이데올리기에 비춘다면, 외국인이 때로는 현실을 보다 세밀하게 파악하는 능력이 있다고 믿고 있다. 왜냐하면 그들은 평범함 속의 특별함을 보다 쉽게 (구별) 인지할 수 있기 때문이다.
이것은 우리 작업의 근거이다. 이제부터 특정한 장소를 선택하고 탐구해보는 범주로 제기된 명제는 '평범함은 다같이 축복되어져야 한다' 이다.
하지만 공유된 평범함을 이해하기 전에, 자국의 평범함 (일반적으로 특별히 주목되지 않는 것) 을 이해하는 것이 중요하다. 이것은 학생들이 자신의 일상을 공유할 대상으로 상대편 국가의 학생을 자신의 집에 초대하는 이유이기도 하다. 그러므로 외국 학생들은 공적장소의 특정한 용도를 결정하는 독특한 문화적 특성과 함께, 일상적 공간의 어떤 연장으로서 공적공간을 인지하는 것을 목표로 해야 한다.
숭실대학교의 요구사항에 따라, 사이트는 최소 1000m^{2}로 선정되어야 했다.
그 사이트들에 대한 우리의 유일한 요구사항은 이 땅들이 물리적인 상황을 넘어서 폭 넓은 의미에서 이해되기를 바랐다. 이러한 폭넓은 이해는 이 장소가 가지는 도시적 현상의 맥락 속에서 정의되어야 할 것이다. 이 전제들에 따라, 다음의 소제들이 제안되었다.

첫 번째 서울 - 알리칸테 워크샵
"Be My City"

A. 이데올로기
경험하는 것은 해석하는것이고, 해석하는 것은 디자인 하는 것이다.
B. 방법론
몇몇의 도시 경험들이 디자인되어질 것이다. 이 경험은 한주의 기간동안 이루어지는데 학생들은 '무엇이 이미 지어졌는가' 에 대해서 배우게 될 것이다. 현지학생의 도움을 받으면서, 우리의 전략은 개인적으로 도시의 현실을 경험하는 것이다.
수동적인 태도는 허락되지 않는다. 그것은 특정한 어느 장소를 보는 것에 국한되는 것이 아니라 모든 경험을 드로잉이나 다른 멀티미디어 등에 의해서 면밀히 기록하는 것이라고 말할 수 있다.
C. 표어:
'사람을 위한 문화 그리고 문화를 위한 사람', 이 표어의 정신은 두가지 다른 시각을 가지고 문화를 이해하는 것이다.
옥스퍼드 사전의 정의에 의하면, 문화는 두 개의 뜻을 가진다.
1. 총체적으로 간주되는 인간 지성의 결과가 예술이나 다른 어떤 형태로 발현된 것
2. 생각, 관습, 특정한 사람이나 집단의 사회적 행동
우리의 작업에 내재된 함의는 정확히 말하면, 단순한 장소의 선택을 넘어서 어떻게 문화의 발현된 현상들이 서로에게 영향을 미치는지에 대한 이해이다.
이 경우에, 두 대지는 '프로그램 되어진' 문화 와 '인식되어진' 문화의 영역을 포함하도록 특징지어진 곳이다. 하지만 동시에, 다른 즉흥적인 문화: 즉 도시적 집합, 사회적으로 지중해적인 것들 또는 이따금씩 숨겨지고 부정된 것들이기도 하다.
하지만 동시에 제시된 두 장소들은 명확한 차이가 있다. 먼저, 첫 번째 대지는 역사에 의해 구축되어진 장소이고, 다른 한 곳은 새롭게 창조될 최적의 순간을 기다리고 있는, 비어있을 뿐만 아니라 역사성이나 정체성마저 없는 공간이다.
D. 장소
site1 'Las Cigarreras' 문화센터
도시패턴속에 삽입된 과거의 담배 공장이고 현재 문화센터로 변화되고 있다.

전시, 콘서트, 영화상영 그리고 설치미술 등 선정된 대지는 '프로그램어진' 문화적인 환경을 제공한다. 하지만 100m 이내에 알리칸테를 대표적으로 상징하는 '베나칸틸'산의 꼭대기에 산타 바바라 (9세기)' 성이 자리잡고 있고 'Las Cigarreras' 를 바로 마주하고 있는 북쪽 산 정상에는 수천의 젊은이 들이 주말 저녁에 현재 가장 쟁점이 되는 '문화적 현상' 인 보테온'을 위해 이 산을 오른다. '보테온' 은 대화, 기억, 일화들을 술과 함께 공유하는 것으로 술집에서 일어나는 상황을 외부공감에서 재생산 하는 행위이다.
우리에게 주어진 도전은 '어떻게 두 순간(낮에 성을 산책하는 것과 저녁에 술병을 들고 서 있는 것)' 을 경험하는 개인적 이해가 예술과 이 장소의 관계를 이해하는 새로운 방법으로 다가올 수 있는가이다.
첫 번째 대지의 이해를 위해 디자인 된 경험은 아래의 세 가지로 구성될 수 있을 것이다.
a. 성에서의 아침
b. 'Las Cigarreras' 에서의 저녁
c. 밤의 'Botellon'

그러므로 이 세 가지 순간의 경험에서 문화센터 디자인을 새로운 방법으로 접근할 수 있을 것인가? 학생들은 그 답을 분명히 얻어낼 것이다.

site2 'La Goteta'
'La Goteta' 는 해변에 가까운 위치에 자리잡은 오래된 도시의 채석장(지금은 쓰이지 않는)이다. 현재는 "유망한" 국회의사당으로의 건축을 기다리고 있다. 이 사이트는 '포스티호엣' 해변과 노동자들이 사는 교외 이 두 지역을 연결하고 시내로 나가는 트램 라인이 필수적으로 지나가야하는 지점이다. 이런 대지의 특성은 의사당에서 일하기 때문에 격식있게 차려입어야 하는 사람들과 해변으로 향하는 수영복과 우산을 들고 있는 거주민들이 섞여있는 역설적인 상황을 가져올 수 있다. 이러한 상황은 하이브리드적인 상황을 분명하게 보여준다.
어떻게 지중해 해안의 여가 경험이 의사당이 혼재하는 상황에 대한 이해에 영향을 미칠 수 있을 것인가?
계획된 경험은 아래 네 가지 활동으로 구성 될 수 있을 것이다.
a. 트램으로 사이트를 지나 '무치비스타' 해변으로 돌아오는 여정
b. 오전 '시에라 그로사' 산으로 하이킹과 피크닉
c. 사이트 방문
d. '포스티호엣' 해변에서의 저녁

그러므로, 이 네 가지 활동의 경험을 통해 문화센터를 디자인하는 방법을 새롭게 할 수 있을 것인가?
학생들은 그 분명히 이 문제를 잘 다룰 수 있을 것이다.

결국, 이런 상이한 상황들이 서로를 부정하는 것처럼 보일지 몰라도, 모든 사건들은 물리적으로 공존하고 있다.

이 워크샵을 위한 각자의 목적은 미래의 디자인이 무엇이든 간에 모든 일들이 공존 할 뿐만아니라 서로에게서 배운다는 것이다.
동시에 우리는 재미, 기쁨, 흥분과 같은 감정이 워크샵을 운영하기 위한 적절한 전략들일 뿐만 아니라 자가 학습이며 디자인 하는 것임을 잊지 말아야 한다. 이것은 정말로 중요한 일이지 않은가?

SEOUL CONCEPT MODEL

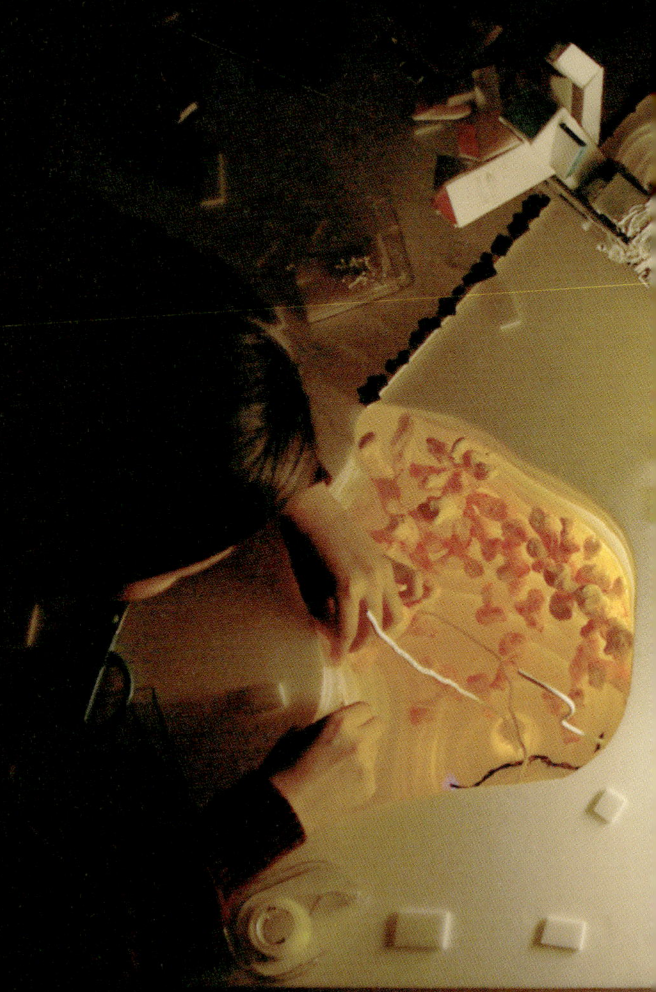

by
Jaekwon Ahn, Chiyoung Park
Mina Kim, Hyemi Bark
Seonghak Cho, Minjae Kim
Juhye Kim, Minjeong Song,
Byungyup Lee, Jihyun Park
Soyoung Kim, Eunmi Park

SEOUL SITES

Alicante + Seoul Workshop 2011

run by Ivan Capdevila + Jungin Kim for the University of Alicante + Soongsil University

Subscribe to feed

Home Visits (Seoul) BE MY CITY! Project Sponsors Alicante Sites **Seoul Sites** Visits (Alicante) Participants Bibliography

Hong-dae

This place was full of art and music with various culture so that people can feel the youth, freedom, passion and creation. But not any more. People who made up the culture of Hong-dae have being converted with the up-coming capitalists and franchises. The locals have being isolated from the visitors in a last few years. Escessive commercialization will lead the place to be devoid of art and culture in the end. Now this place feels keenly the necessity of being filled with essential art and culture as like the origin of the place.

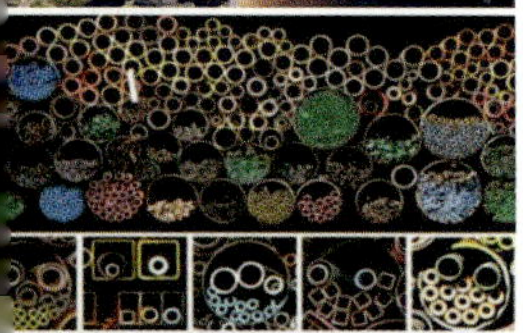

Mullae Art Village

Several artisit want to live with welders in Mullae.
We need to build the city where these two groups can coexist and live together.

Tunnel in Sangdo-dong

Seoul is mountainous city. First, I noticed the residences on top of tunnel. The city has a lot of people and many cars, which casued traffic congestion. In order to overcome this mountain, condition and traffic ams we have to drill tunnels. You know Koreans are the best drillers.

Song-do New City

Songdo is a reclaimed land. By being a man-made tabula-lasa, the future city is a dream country for all failed modernists. What you see is the Radiant City resurrected with the currency of global capital. Symbol of "delirious Seoul"and epitome of "koreanized modernization". The devil is back.

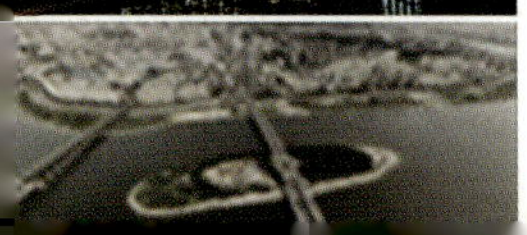

Nodle Island

Seoul is city of bridges. They connected the north of the Han-river and the south. Among them, the Han-river Grand Bridge is a famous one, because it straddles both banks over passing Nodle Island. Ill de la Cite. But it is a nobody's land, a blank space. So I invite you to propose some events.

Haebang-chon in Yongsan-dong

Haebangchon is close by Yonsan Grrison-head quarters. The area of army has not Korean address but also American address in California. Harsh crashing culture has been happened. Then, nowdays a unique culture of regional culture has been formed surrounding the army.

Se-un Sang-ga

Se-un Sang-ga which was put in center of Seoul in 1966 is literally a monster. As time goes on, city has sucked the monster's energy and expanded their own boundary which is like indepenedent nation. What happened here.

Dongdaemun Market

'Dongdaemun market' is very big scale market and mix of several style of market. nevertheless 'dongdaenum market' seems like a one the reason that street was used for space as well as path. many event are happened on street of dongdaemun market so street was constructing network in dongdaemun(urban) street used for people's life space and street become unique space in urban.

Itaewon

You can see the picture taken in Itaewon. Picture on the right side is a mosque. In front of it, yoou encounter Gay bar strip. Both of which are still alien to ordinary experience in Seoul, but they exist and everybody feels freedom here.

Apartment Comlpex in Ichon-dong

Seoul is full of Apartments. You see the apartment complex all around Seoul. You can't understand Korean culture without understanding on apartment. All of scenes and frames you see always will be with apartment.

Sechon Village

When I walk down the streets of Sechon, I can feel the small traces of history that have been engraved into the slowly modernizing surroundings. The mixture of traditional houses and village-like atmosphere with the formation of new buildings that pay tribute to the past leave a deep impression of home. The changes of this town are of its own manner and speed, unlike the ever increasing amendments of the surrounding metropolitan Seoul area. However, the changes of the future are inevitable, so the question is what kind of change is best for the future of Sechon: one of historicism, modernism, both, or something else.

Teheranno

At night, "Teheranno-front-people" go to meet "Teheranno-backside-people". "Teheranno" of Seoul is most rich and big IT Street. There are have Six-car road, and so many high-rise building. So some people that don't know there think "Teheranno" is to clean and bright. But ! Every evening there start to change their face. First, so many sexual add paper increase all of the Teheranno-street. Second some people tout to people of workers or anyone And then Teheranno-main street is so empty, but backside ofTeheranno-street have activition of sexual and so many add paper.

ALICANTE SITES

Alicante + Seoul Workshop 2011

un by Ivan Capdevila + Jungin Kim for the University of Alicante + Soongsil University

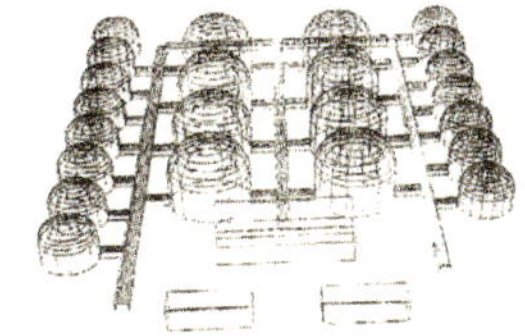

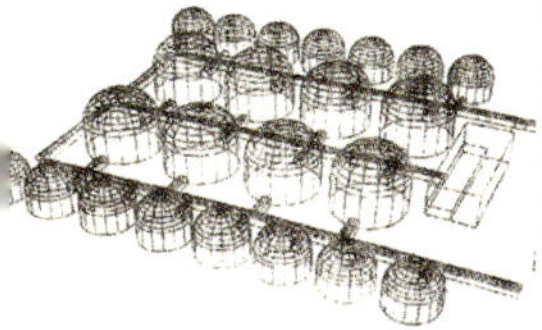

La Sangueta

1875_ the Deutsch&Cocompany set up an industrial complex to refine crude oil in the Serra Grossa's border, next to the quay where they received oil.1900_ at the beginning of the XX century the refinery supplies 55% of the Spanish oil market.1915_ a new railway line connects the refinery to Alicante's harbor, so the quay is no longer necessary.1937_ the huge demand of oil generated by the Spanish Civil War and the Second World Wide War causes the need of extension of the refinery. Excavations began at the end of the Civil War and finished at the beginning of the 50's.2006_ buildings and outdoor tanks were demolished in anticipation of the tram works.

La Sangueta는 오일을 들이는 부두 옆으로, 네덜란드회사가 천연오일을 정제할 수 있는 산업단지를 조성하면서 시작한다. 1900년대로 들어서면서 이 곳에서 정제되는 오일은 스페인 오일 마켓의 55%를 차지할 정도로 비중이 커졌다. 그 후, 철로가 들어서면서 이전에 오일을 들이는 유일한 통로였던 부두는 더 이상 쓸모 없게 된다. 1937년, 스페인 시민전쟁과 2차 세계대전을 거치는 동안 막대한 양의 오일의 수요가 발생하게 된다. 그 여파로 La Sangueta의 내부는 계속해서 확장을 하였다. 2006년 이 곳을 지나는 트램이 생기면서, 내부구조물과 외부 유류탱크는 철거되며 현재에 있다.

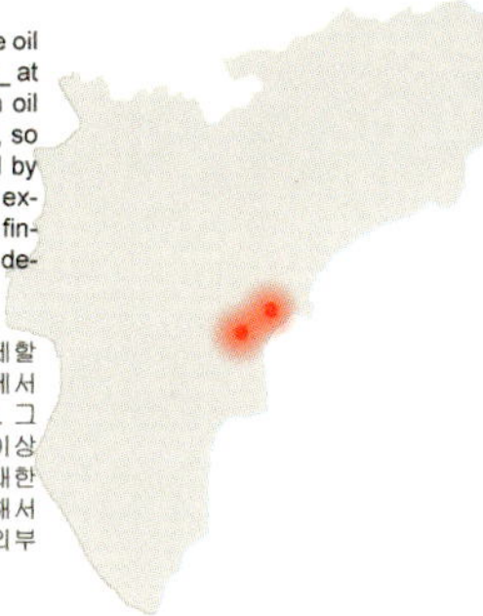

Las Cigarreras

At first it was a church and a religious house in 1750. That kind of buildings were been searched by the company to create the factory. Tabacco factory was created Alicante harbour city in 1804. A part of the building was destroyed by fire in 1840. In the early 1900s, new machinery was introduced and surface of the factory was increased in 1936. Factory started rebuilding in 1960 and finished in 1970. The new factory opened, as cultural center, to outside the city in 2011. Throughout history, the program of the building has changed constantly. Before the Industrial Revolution, when religion was the most important factor, it was used as a religious building. Then, during the Revolution era, it was used as a factory and currently, it is being used as a culture center. So then, what next?

La Cigarreras는 종교건물에서 기원한다.Bishop(카톨릭의 주교)의 거주지였고, 종교적 믿음을 통한 금욕의 장소가 그 출발점이다.비숍은 지역의 오갈 데없는 가난한 여성들이나 매춘부들을 모아 이곳에서 살도록 해주었다. 그 후,1801년에 공장을 설립하려는 회사에 의해 채택되었다.처음은 빌딩의 일부분이 사용되었지만, 점차 전체가 담배공장의 용도로 완성되었다.1900년대 산업혁명이 시작되면서, 새로운 공정에 적절한 장소를 찾아 공장은 이전하게 되고, 버려졌던 이 건물은 2011년 문화센터로 다시 태어나게 된다.Alicante 공장에서 일어난 이 모든 것들은 스페인에서의 산업혁명의 발전사를 극명하게 보여주는 예이다. 그리고 그 당시 Alicante의 경제와 사회의 단면이기도 하다. 스페인에서의 민주화의 도입과 함께 담배 산업의 전매는 민영화로 넘어가게 되고, 스페인 공장들의 많은 폐업을 잇게 하는 등 산업에 깊이 영향을 미쳤다.

Alicante

알리칸테 주는 스페인 발렌시아 지방의 주이며 지중해연안의 항구도시이다.
주도는 알리칸테이며, 인구는 1,891,477명(2008년)이고, 인구밀도는 325.19명/km²으로 스페인 전체 인구의 4.1%를 차지한다.
알리칸테는 천년역사의 도시이며 과거 남겨진 유적들 뒤에 남겨진 골목길들은 이러한 사실을 증명해 주고 있다.

어떤 특별한 이벤트를 기대하지 않아도 스페인 광장 그 자체가 이벤트의 연속이다.
Never containing any particular purpose, The Plaza, itself was a places for continuous events.

Privacy in the public, Toledo

중세도시 톨레도, 시간이 정지한 듯한 도시의 풍경을 끝없이 쫓아다니게 된다.
Toledo, Chasing after the frozen time.

사람들은 항상 어딘가 머물러 있고,
우리 또한 자연스럽게 머물고 있다.
And we want to stay somewhere in-between.

해를 찾아 나온 사람, 해를 피해 나온 사람 모두에게 지중해의 해변은 풍요롭다.
The Mediterranean sun is for everyone.

해가 떨어지는 각도와 선글라스의 굴곡이 이루어낸 저녁 6시경의 풍경
Alicante's sunset reflects in Eunmi's Sunglasses.

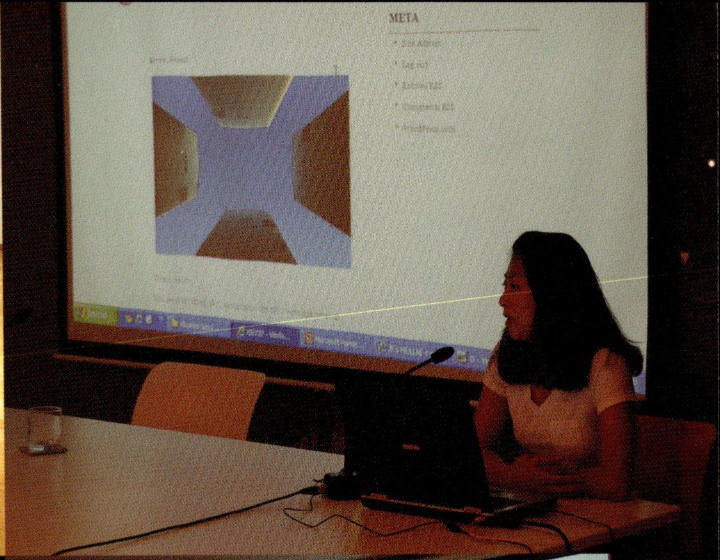

알리칸테의 지역건축협회에서 서울을 소개하는 한국 학생들
Korean students are presenting their works at Colegio Territorial de Arquitectos de Alicante

프로젝트에 대한 알리칸테 지역 방송국 "I TV"와의 인터뷰
Interview with Alicante local tv station, "I TV", about the project.

CONFERENCIA:
The city of soul: Seoul
JUNGIN KIM
Architecture Professor Soongsil University. Seoul, South Korea
Lugar:Salon de Actos P4 Viernes 7 octubre 13.30h
ORGANIZA Y COORDINA:
Área de Proyectos Arquitectónicos. Universidad de Alicante
Universitat d'Alacant
Universidad de Alicante
con el patrocinio de:
LEVANTINA
THE NATURAL STONE COMPANY

CULTURAL MIXING: ALICANTE REDISCOVERED

Jungin Kim

Our first encounter with Europe was a more playful and a recreational one. Instead of cramming intensely the ideas of conventional practice of architectural pedagogy within confined studio spaces, we choose to meet open airs and spontaneous encounters. Our choice thus engaged with urban phenomena manifested in Alicante's urban places as well as the city's lasting impressions which intuitively embraced by each design team as an individual narrative for telling the culturally different European urban places.

Seonghak and Minjae were interested in the culture of Spanish youth, especially their enigmatic situation being pushed out to the border of established old system and caught in the middle of the society's impending crisis. Botellon is viewed to them as a way out for the Spanish youth and as a festival that somehow need to be accommodated through urban intervention. As they strongly sympathized sympathy with the condition that the Spanish youth are resistive to the impasse imposed by the older generation, and express this frustration through the collective drinkingactions in the urban places, the team used the existing tram line and expanded it to create tram-line fabrics of La Sangueta whose networks accommodate the Botellon culture and locate themselves within. In this juncture, Korean any-where any-time delivery system as the vital part of Seoul's urban life is incorporated into the ramified lines of trams and becomes a methodological device in which Botellon is played out in nomadic and flexible patterns.

Having visited Las Cigarreras and being emotionally responsive to the dynamic scale and sequential spaces within the historical but abandoned building, Juhye and Minjung developed an interesting notion between density and lifestyle. Their interests dwelled on the bustling urban density in Seoul and how these qualities are introduced to the building. By filling the background spaces of the building and making them hybrid spaces with colorful ambience, Alicante's light becomes a main design medium that proposed programmed boxes which the large portion of public spaces in-between becomes an integrated whole. In this animated spaces of multi-variant lights they were trying to combine the liveliness and dynamism of Asian cities with the Piranesian space that is sit idle under the Mediterranean sun. Their effort was to design not the thing as expressed in specific uses, but between things as spaces are open to any activity.

Jihyun and Byungyup approached Las Cigarreras somewhat differently from Juhye and Minjung by tracing a gendered and sexualized notion of the place. In an effort to reveal the particular use and (her) story of the building,the team adopted to this Spanish building the gendered condition in Seoul's voyeuristic spaces, most dramatic one being the JimJilBang. JimJilBang is a Korean bathing house that gained rapid popularity when South Korea underwent economic crisis in 1997. Here often hierarchical, formalized, and custom-ridden behavioral protocols in the Korean society are evaporated with the warmth of bathing environment. Private gestures like hug, sleeping, and kissing are permitted in the rooms where each different uses for JimJil activity (literally steaming action or sauna) are accommodated while bathers are dressed in uniformed robes. Korean JimJilBang and male gazes are problematized spatially, but these are chosen to be cultural programs in order to meet the abandoned structures in Alicante.

When SoYoung and Unmi see the empty and bleak open space of La Sangueta, they tried to bring liveliness of Seoul and its street environments activated by "PoJangMaCha", a name for Korean street vendor. These mobile and personal stalls are extremely adaptable to pedestrian traffics and solicit passersby displaying small personal paraphernalia, accessories, and many versions of spicy and sweet Korean snacks.

In the manner of mixing Korean artifacts of urban culture into Alicante's sites, Mina and Hyaemi was interested in applying to Las Cigarreras the popular pre-school system in Seoul that is fundamental to everyday life of Korean parents. While their parents are at work, children can play themselves in the culturally hybrid spaces that contain various element of street cultures both in Seoul and Alicante. Likewise Jaekwon and Chiyoung were also trying hard to introduce into La Sangueta the Koreanized Fish Market of Noryangjin and NoraeBang, the Korean singing room. Their semiotic grasp of the site as 'dual harbor' anticipates the revitalization of Alicante's largest empty plane from these culturally unfamiliar, but intriguing programs.

MEZCLA CULTURAL: ALICANTE REDESCUBIERTO

Jungin Kim

Nuestro primer encuentro con Europa fue más bien uno animado y recreativo. En lugar de atiborrarnos de ideas propias de una práctica más convencional de la pedagogía arquitectónica en el confinamiento del aula, preferimos el aire exterior y los encuentros espontáneos. Nuestra elección, así pues, tenía que ver con fenómenos urbanos que se manifestaban en determinados lugares urbanos de Alicante. La impresión intuitiva más duradera de la ciudad por parte de cada equipo construiría una narrativa individual cuyo objetivo sería el de evidenciar la condición urbana europea, tan culturalmente diferente.

Seonghak y Minjae se interesan por la cultura de la juventud española, especialmente su enigmática situación llevada al límite de lo establecido por el sistema y pillados en medio de una crisis social desalentadora. La juventud ve el "botellón" como una salida y como un festival que, de alguna manera, necesita acomodarse mediante una intervención urbana. Dada su gran simpatía por esta condición de la juventud española de resistirse a las imposiciones de generaciones anteriores y expresar sus frustraciones a través de beber colectivamente en lugares urbanos, el equipo utiliza la línea existente del tranvía para expandirla y crear nuevos tejidos de vías en La Sangueta cuya red acomoda la cultura del Botellón. Esta red, a su vez, se integra en esta cultura.

Tras haber visitado Las Cigarreras y respondido emocionalmente a la escala dinámica y secuencia de espacios dentro de sus edificios históricos pero abandonados, Juhye y Minjung desarrollan la interesante noción de densidad y estilo de vida. Sus intereses habitan dentro de la bulliciosa densidad urbana de Seúl y cómo estas cualidades pueden introducirse en el edificio. Llenando los espacios del fondo del edificio y convirtiéndolos en espacios híbridos de un ambiente colorido, la luz de Alicante se convierte en un medio principal de diseño. Cajas programáticas y una gran parte de espacios públicos intermedios se convierten en un todo integrado. En estos espacios animados de luces múltiples intentan combinar la viveza asiática y dinamismo con el espacio Piranesiano sin uso bajo el sol mediterráneo. Su esfuerzo es el de proyectar no desde usos específicos sino entre aquellos lugares abiertos a cualquier actividad.

Jihyun y Byungyup se aproximan a Las Cigarreras de manera diferente a Juhye y Minjung definiendo una noción del lugar más relacionada con lo sexual y el género. En su esfuerzo por revelar el uso específico y su particular historia del edificio, el equipo adapta a este edificio 'español' la condición de género del espacio voyeur más dramático de Seúl: el 'JimJilBang'. El 'Jimjilbang' es un lugar de baño típicamente coreano que ganó un rápida popularidad cuando Corea del Sur atravesó una fuerte crisis económica en 1997. Aquí, los tan a menudo protocolos formales, de jerarquías, de comportamientos encorsetados intrínsecos de la sociedad coreana se evaporan con el calor del entorno acuático. Gestos del ámbito privado como abrazarse, dormir y besarse están permitidos en habitaciones grandes y pequeñas donde cada uso diferente en la actividad 'JimJil' (literalmente 'acción de vapor' o sauna) se acomoda con bañistas vestidos con un manto uniforme. El 'JimJilBang' coreano y las miradas masculinas se problematizan espacialmente, aunque se acaban conformando como un programa cultural para hacer frente a estructuras abandonadas.

Cuando ven el desolado y vacío abierto de La Sangueta, SoYoung y Unmi intenta traer algo de la viveza de Seúl y su entorno de calles activadas por los 'PoJangMaCha', nombre de los vendedores callejeros coreanos. Estos vendedores móviles y personales son extremadamente adaptables al trífico peatonal importunando a los viandantes con todo tipo de parafernalia personal, accesorios y muchas versiones de típicos aperitivos coreanos picantes y dulces.

Buscando la mezcla de artefactos coreanos de cultura urbana con los solares alicantinos, Mina y Hyaemi se interesan en importar a Las Cigarreras el popular sistema de preescolar en Seúl enraizado en la vida cotidiana de los padres coreanos. Mientras sus padres están en el trabajo, los niños pueden jugar por sí mismos en espacios culturales híbridos que contienen varios elementos de culturas urbanas tanto de Seúl como Alicante.

Del mismo modo, Jaekwon y Chiyoung intentan introducir en La Sangueta el "coreanizado" Mercado del Pescado y el Noraebang (equivalente coreano del karaoke). La consideración semiótica del sitio como 'puerto' anticipa una nueva revitalización de la planicie más vasta de Alicante desde estos programas culturalmente no muy familiares pero intrigantes a la vez.

숭실
반점

BOTELLONCITY

Botellon is a social activity among Spain's youth, who gather in public areas to consume alcohol. But it's illegal.

01

PARTICIPANTS

02

URBAN EXPERIENCE

Always drunk alcohol in Spain

Question : Why?

But, No drunk in beach once

At night, we drunk in the beach again

04

Why did not we drink alcohol in the beach during the daytime?

BOTELLON

A social activity among Spain's youth, who gather in public areas to consume alcohol as an alternative to going to a bar or club. But it's illegal. The first reason of doing botellon is a economic situation of the young people

05

Ordinary time table of Spanish

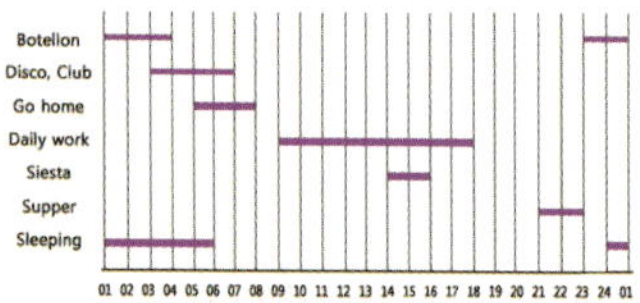

06

Botellon Participants

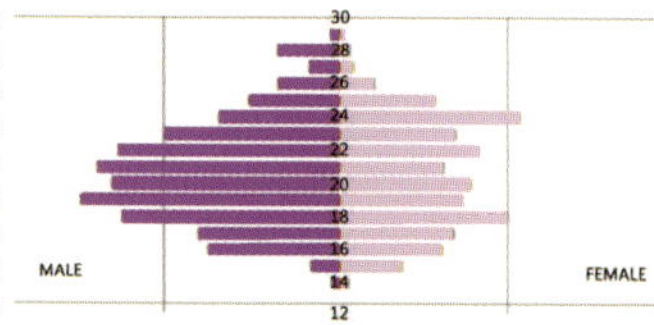

Botellon : Social problem

Noise Dirt Moral issues Healt

08

So we dicided to make a

CITY FOR BOTELLON

09

What needs for Botellon

In order to have botellon, we need some elements. Botellon is like natural activity so if we want to make artificial botellon, we have to analyze botellon.

Pattern of Botellon

FRIENDS

First get drink and then meet the group

RAVES

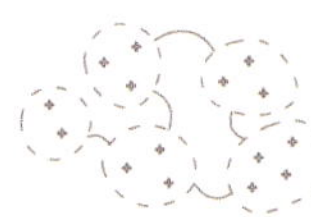

Get drink and meet people on the way

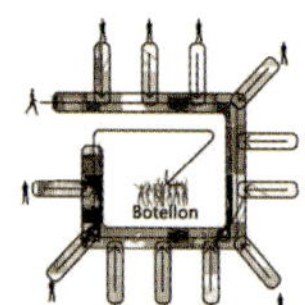

Botellon can be categorized to 2 pattern basically., group and solo. Futhermore botellon types can be characterized by where it happens

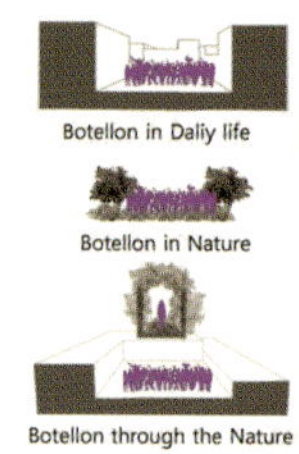

11

LA SANGUETA

Attractive Void for Botellon

12

Site condition

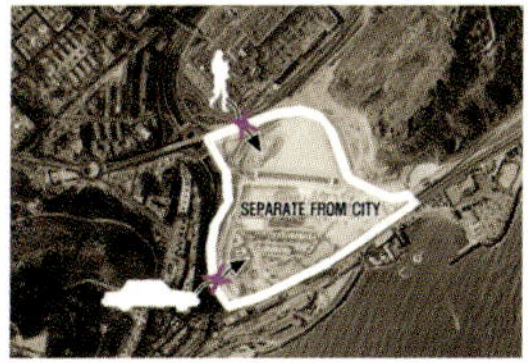

No walking, No car

13

How to approach

TRAM

The Alicante Metropolitan Tram operates in the Spanish city of Alicante and its Surrounding area.

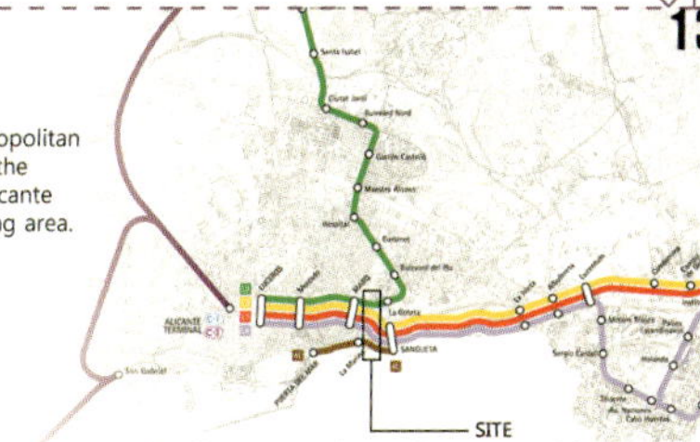

One line stop at the La Sangueta and four lines pass by this place. Here is very important pivot of tram.

14

Delivery Tram

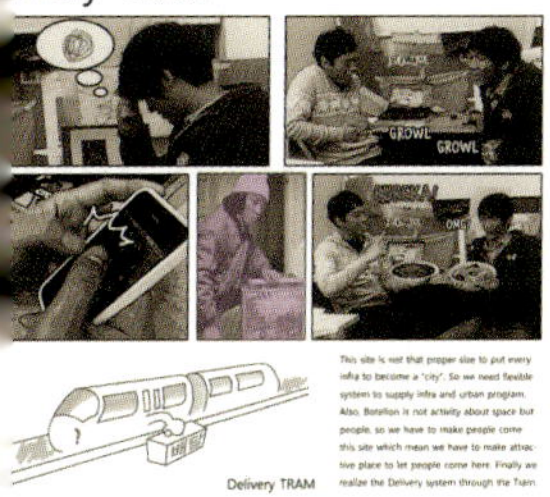

Delivery TRAM

This site is not that proper size to put every infra to become a "city". So we need flexible system to supply infra and urban program. Also, Botellon is not activity about space but people, so we have to make people come this site which mean we have to make attractive place to let people come here. Finally we realize the Delivery system through the Tram.

15

Design Development

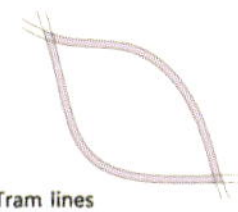

1. Tram lines

2. Mass shapes follow tram lines

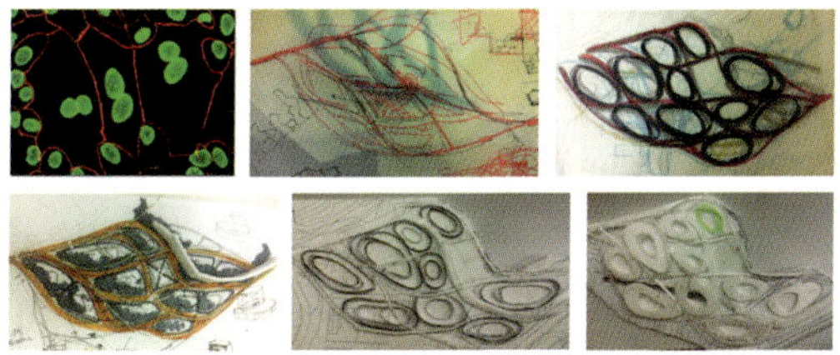

16

New TRAM system

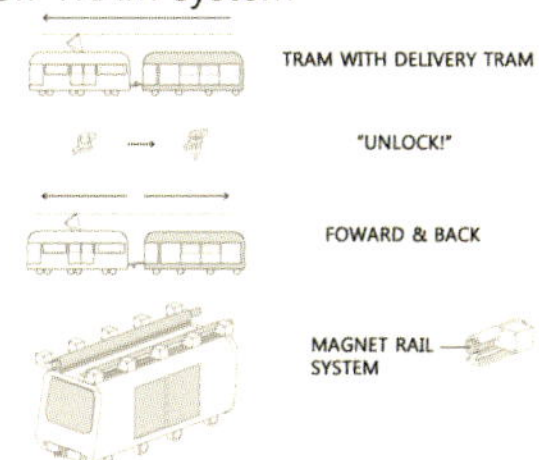

17

Tram's Program

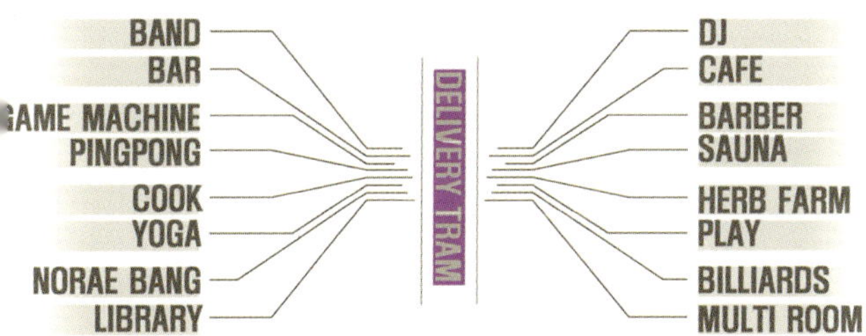

18

Structure

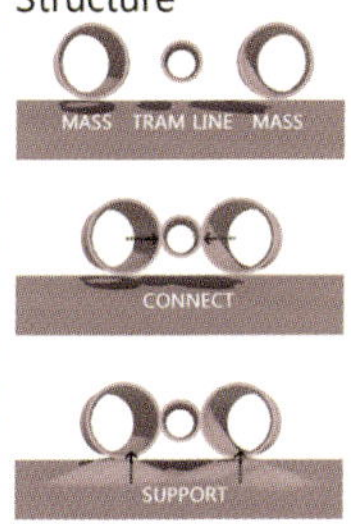

Each masses can be supported by ground shape and tram lines can be supported by each masses. All structures are supported by each others

19

Delivery System

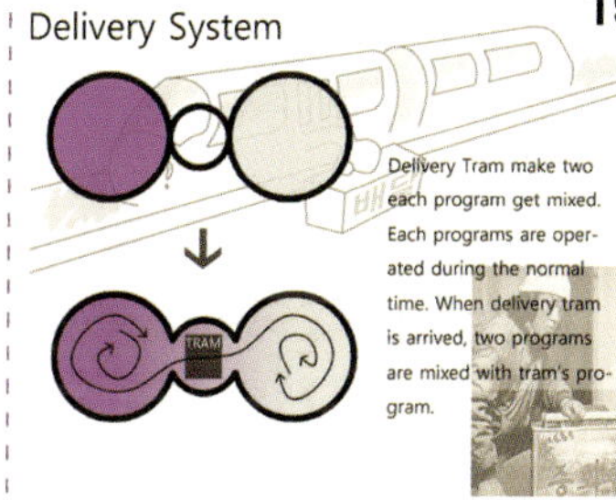

Delivery Tram make two each program get mixed. Each programs are operated during the normal time. When delivery tram is arrived, two programs are mixed with tram's program.

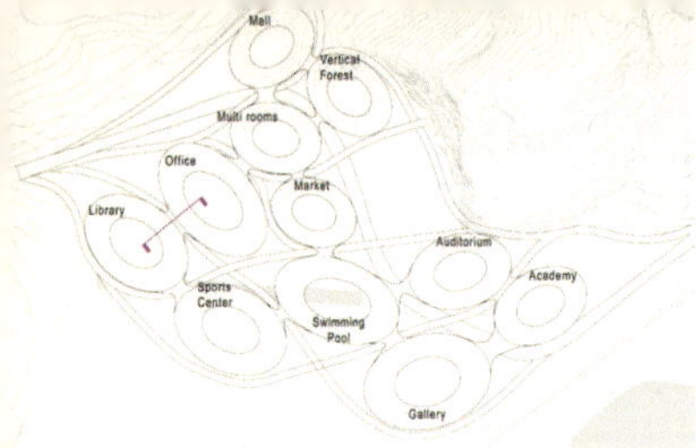

이 대지는 도시로 디자인해서 모든 인프라들을 담기에는 적당한 규모가 아니다.
그러므로 도시 프로그램과 인프라를 제공하기 위한 유연한 시스템이 필요하다.
보테욘은 공간보다는 사람에 치중된 행위이기 때문에 대지에 사람을 끌어드리는 것이 우선과제이다.
쉽게 말해 사람이 올만한 매력적인 장소를 만들어야 한다.
결국 우리는 인프라를 배달하는 트램시스템이 해결책이 되리라 생각했다.

This site is not the proper size to put every infrastructure of a "city".
We need a flexible system to supply infrastructure and an urban environment.
Also, Botellon is not an activity about space but people. We have to make people come to this site which means we have to make an attractive place to let people come here.
Finally we realize that a Tram will be the best solution for the transportation system.

01

02 URBAN PROGRAM DELIVERED

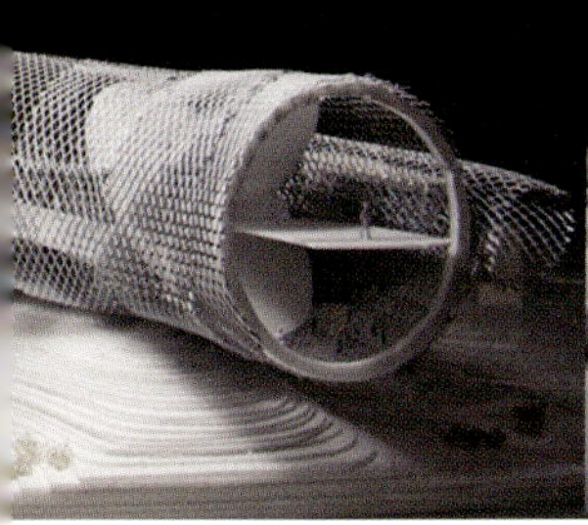

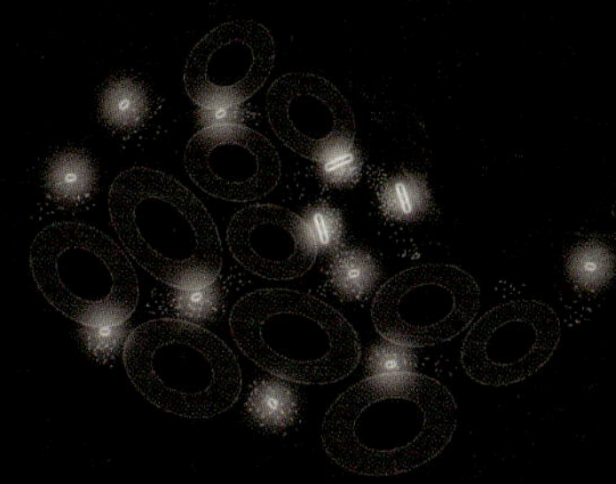

3

04 BOTELLON PROGRAM DELIVERED

We are making “the Den-city”

DENSE/ COMPACT/ TIGHT

a lot of signs and people of myeong-dong street

San juan beach of Alicant

LOOSE/ RELA

여유로운, 그러나 자칫 지겨워질 수 있는 알리칸테.
그리고 집약되고 늘 활기 넘치는 서울 도시의 모습이 섞인다.
프로그램이 있는 공간과 그 외의 공간으로 우리가 본 알리칸테와 서울이 나뉘는 듯 하지만 곧 알리칸테의 뜨거운 태양, 자연의 빛과 서울의 다양한 간판과 같은 인공적인 빛이 하나의 공간을 완성시킨다.

We think of Alicante as a very relaxed space and Seoul as a busy, high-density city. These can be mixed. This space is divided into program space and negative space. We can control the natural light and the artificial light. Finally these two different lights and spaces will mix.

Ju-hye Kim/ Min-jeong song

LET'S GO!
SEOUL
ALICANTE
Gangnam
Itaewon
Myeong dong
CHOOSE THE PLACE
CHOOSE THE PROGRAM
GO INSIDE
OWN AREA
OWN PROGRAM
SEAT ANYWHERE
SPACE
RELAX
DENSE
PROGRAM BOX
RELAXED SPACE
INTERACTION
Space of existing buildings
Space for programs
Forming negative space
Interaction between these spaces

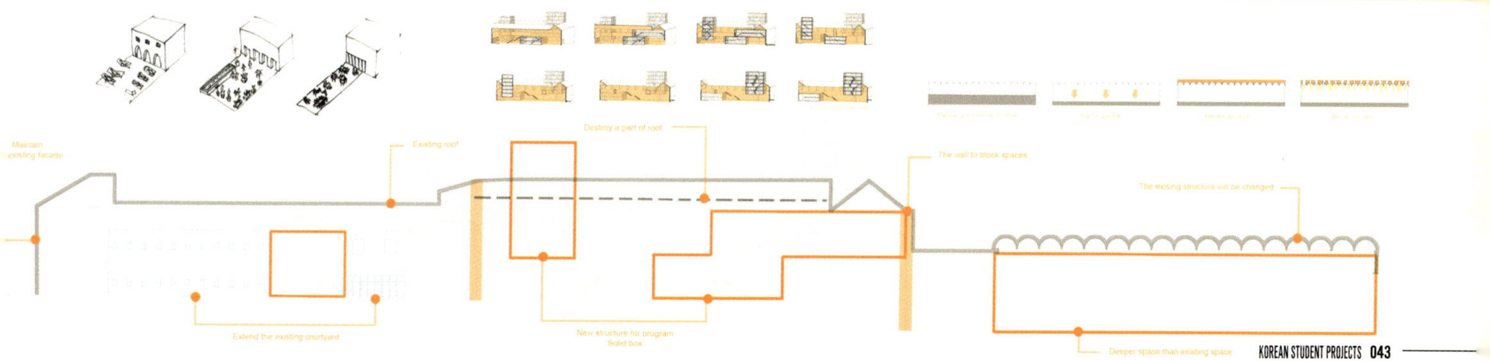

Maintain existing facade
Existing roof
Destroy a part of roof
The wall to block spaces
The existing structure will be changed
Extend the existing courtyard
New structure for program
Solid box
Deeper space than existing space

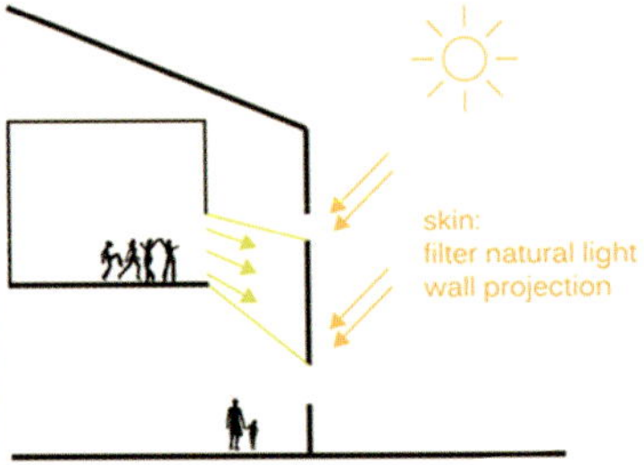

우리는 알리칸테의 자연적인 빛과 서울의 인공적인 빛을 이용해 프로젝트를 진행하였다. 서울을 함축시킨 솔리드 박스에서 뿜어나오는 인공적인 빛과 보이드 공간에 자연적으로 비춰지는 알리칸테의 빛. 이 둘이 서로 상호교환하며 자연스럽게 믹스된다.
여기서 우리가 주목해야할 것은 보이드 공간이 솔리드 박스의 스크린 역할을 함으로써 그림자와 빛의 변화로 인해 사람들의 행위를 추측하게 만드는 호기심과 상상력을 자극하는 공간을 만들었다는 것이다.

The most important thing is the light and shadow. We can control the light with the structured skin, solid boxes and voids. The structured skin works as a screen and the solid boxes have an individual program with their own colors.
The natural light goes through the structured skin and the artificial light shines through the solid boxes. This way you can guess what actions take place in the boxes with the reflected light. So in the void space two different lights will mix.

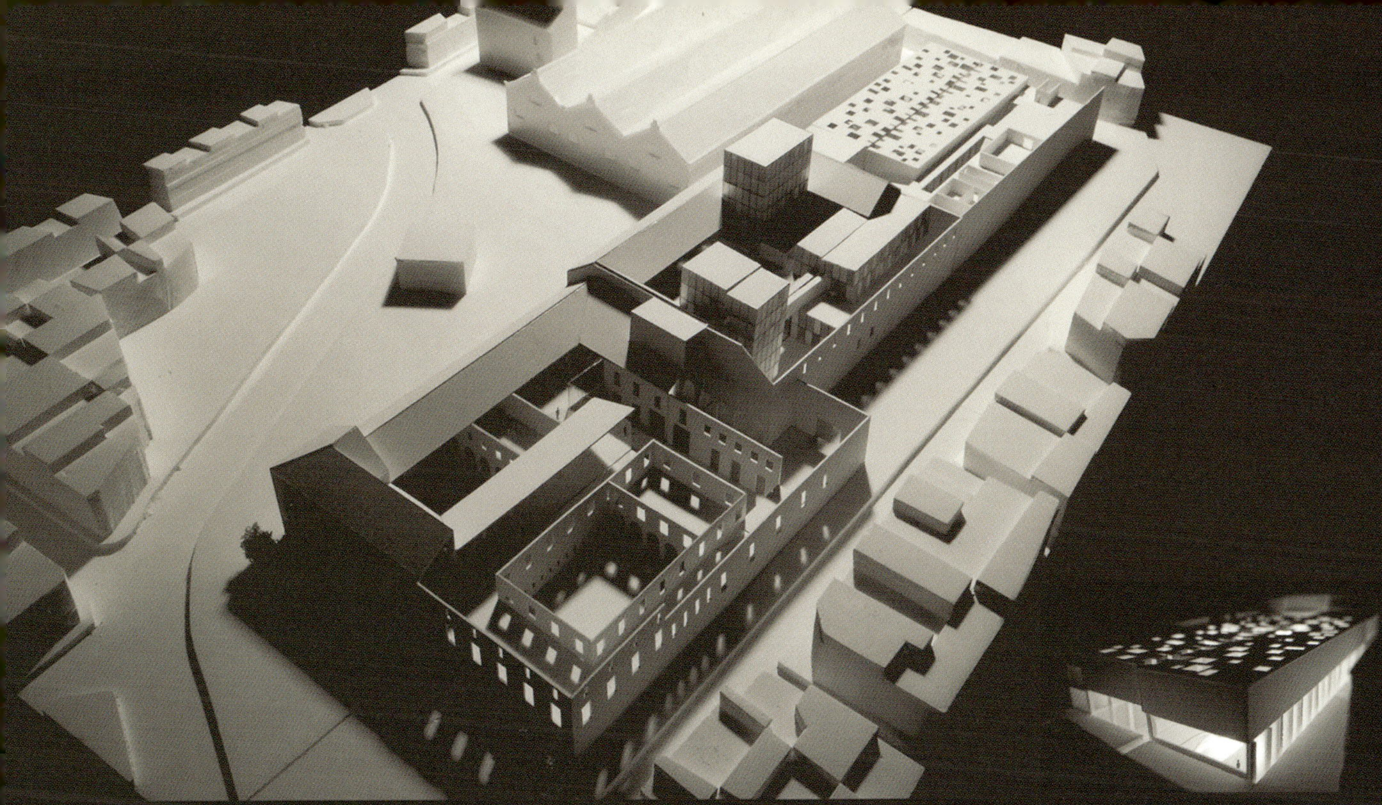

JIMJILBANG in SPAIN

A Jimjilbang (찜질방) is a large, gender-segregated public bathhouse in Korea.

LA CIGARRERAS

폐쇄적인 공간, 그 속에서 담배를 마는 여성 노동자들. 라시가레라에 머물렀던 시간 동안 아주 은밀하게 그 공간을 엿보아야 했다. 그 먹먹한 분위기와 여성 노동자들만의 공간이었다는 이야기는 공간에서의 상상력을 끝없이 자극했다. 우리는 어떠한 방식으로도 이 곳에서의 경험을 잃고 싶지 않았고 그 방법으로 스스로를 관음증이라 이름 붙였다. 또한, 이 버려진 담배공장을 알리칸테의 도시 흐름으로 재결합시키고 싶었다. 우리는 한국적 커뮤니티라고 생각하는 '찜질방'을 제안했다.

One of the most memorable experiences we had in La Cigarreras was seeing women workers in that really closed factory space. Its atmosphere and history make us imagine some intimate situations for that site. After our trip, we didn't want to lose the feeling of the unique experiences and decided to name it Voyeurism. Also, another mission was how to link this closed building with the urban context and open this abandoned space to the Alicante people. Finally, we came up with the Jjimjilbang idea for public space, based on Korean culture.

Experience Model

Lee Byung Yup

Park Ji Hyun

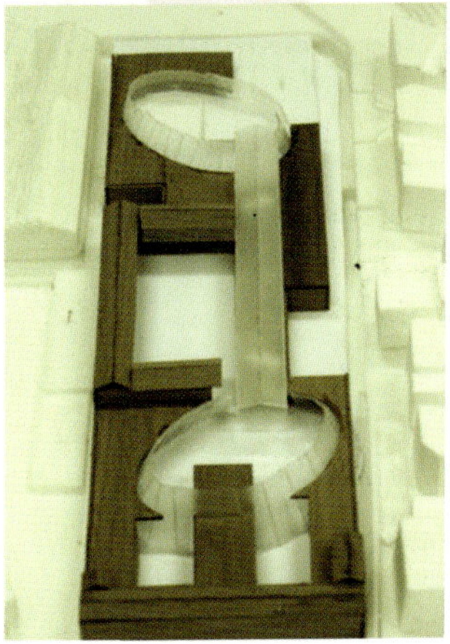

TEMPERATURE & VOYEURISM

찜질방은 온도를 통해 프로그램이 구분되며, 중심이 되는 넓은 마루와 같은 공간과 이어지게 된다. 찜질방에 존재하는 온도차를 이용해서 관음증을 어떻게 유발할지에 집중하였다.

We divide the spaces in a Jjimjilbang by temperature. A central wide plaza connects all of the places with different temperatures. That's why we focused on it when making the voyeuristic situation like a Jjimjilbang.

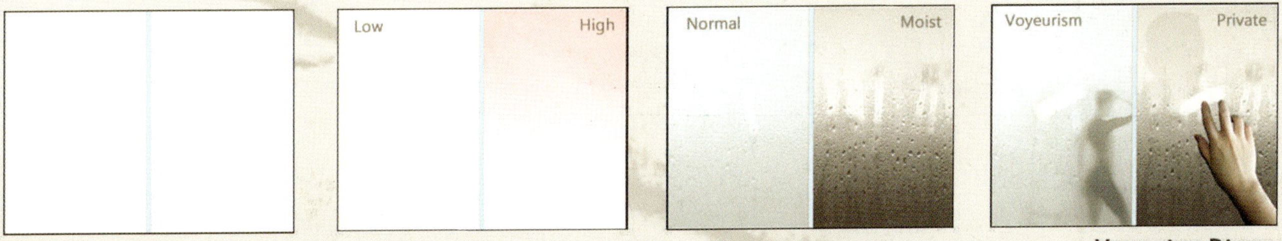

Voyeurism Diagram

Mass Diagram

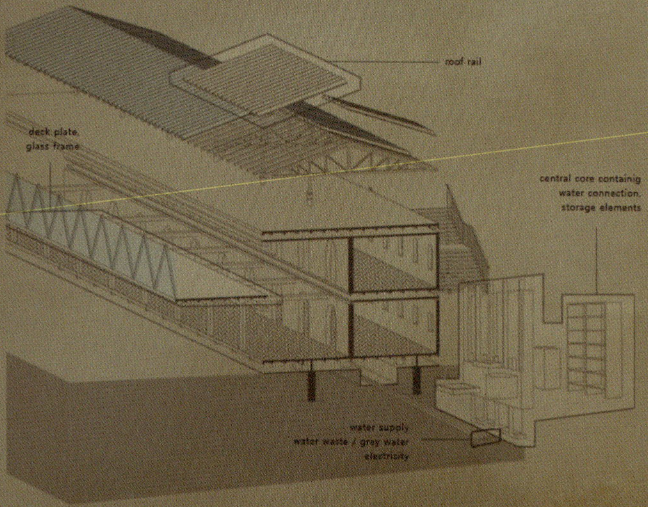

타원형태의 유리메스는 빽빽한 공장 건물의 틈을 만들어 새로운 관계를 형성한다.
Ellipse shaped glass are inserted into the original factory building transforming it into a new space.

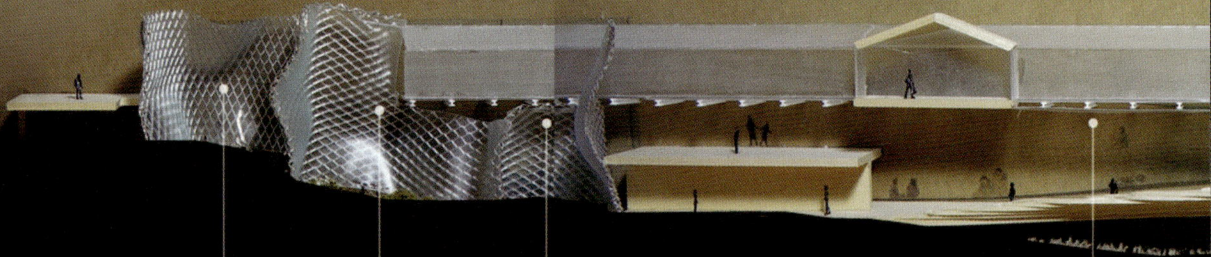

Existing Situation

Jimjilbang Situation

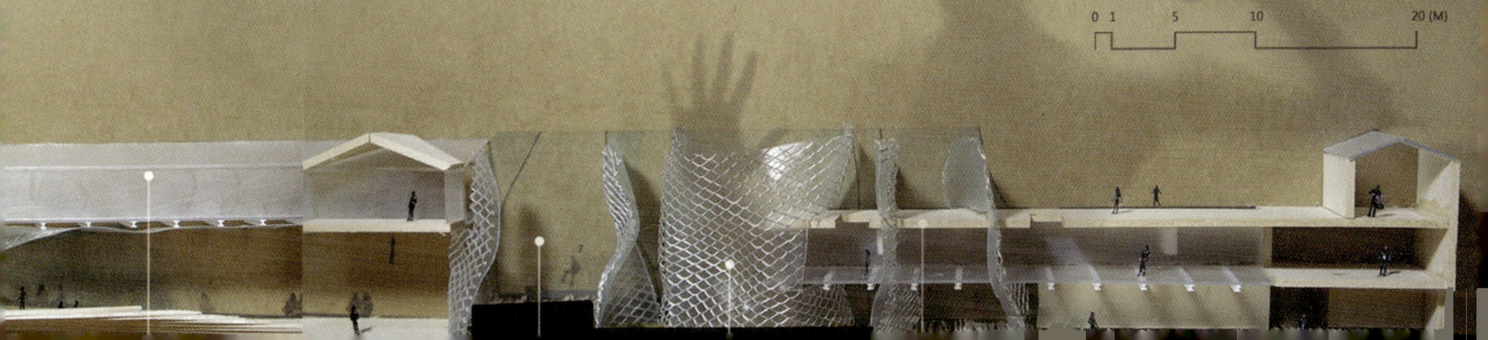

BETTER CITY BY CHILDREN

친구들은 말했다. 스페인의 경제위기에 대해,
그래서 우리는 생각했다. '우리가 그들을 위해 뭘 해줄 수 있는가'
우리 모두는 안다. 미래의 우리 아이들이 해답이라고.

FRIENDS TOLD ME ABOUT THE CRISIS OF SPAIN.
SO WE THINK 'WHAT CAN WE DO FOR THEM?'
AND ALL OF US KNOW. THE CHILDREN OF THE FUTURE ARE THE KEY.

I'M BARK HYEMEE
작업을 맡은 박혜미
AND KIM MINA.
그리고 김미나 입니다
FROM NOW OUR MASCOT, ALICANTE BABY WILL EXPLAIN OUR PROJECT.
HI! I'M ALICANTE BABY! I'LL INTRODUCE THIS PROJ.
안녕하세요! 알리칸테 베이비에요. 지금부터 이 프로젝트에 대해 설명해드릴게요
SE VENDE ó SE ALQUILA 965 13 01 66
SE VENDE 661 231 015 965 104 202
ALQUILA 965 20 63 88
WE HAVE SEEN SEVERAL SELLING HOUSES IN ALICANTE
저희는 알리칸테에서 집을 내놓는다는 광고를 여럿 보았어요
AND CHICHI SAID,
그리고 알레한드로가 얘기해 주었지요.
SPAIN IS IN CRISIS
스페인은 지금 경제
"WE KNEW ABOUT SHORTER BUDGET FOR CHILDREN. AND THERE ARE FEW DAY CARE CENTER IN ALICANTE".
"우리는 알리칸테에서 아이들의 예산이 삭감되고, 공공보육 시설도 부족하며"
"THE EMIGRATED HAD LOST THEIR TRADITIONAL CULTURE"
"이주민들이 그들의 문화를 잃어가고 있고"
"THE OLD SPENT THEIR TIME DOING NOTHING".
"노인들은 무료한 시간을 보내는걸 알았어요"
"AND ESPECIALLY ARTISTS SIT ON THE STREET WITHOUT SPACES FOR EXHIBITION AS CORRUPTION OF THE CITY"
"특히 예술가들은 알리칸테의 부정부패로 전시공간을 구하지 못해 거리에 앉아있었지요.
WHAT CAN WE DO FOR THEM?
우리는 생각했어요. 그들을 위해 무엇을 해줄 수 있을까?
AT FIRST WE MUST CHANGE THE RECOGNITION ABOUT THEM.
먼저 그들에 대한 인식부터 바꾸어야 했어요
SO WE CHOOSE LAS CIGARRERAS !!
그래서 우리는 Las Cigarreras를 선택했지요!
AT THE PAST LAS CIGARRERAS HAD GIVED SOCIAL POSITION TO WOMEN
과거 담배공장이었던 Las Cigarreras는 여성들의 사회적 지위 올려주었던 곳이에요.
BY CHANGING COMMON PEOPLE'S AWARENESS
일반인들의 인식을 바꿔놓음으로써 말이죠
FROM NOW, IT'LL PROVIDE THE SPACE FOR THEM AND CHAGE AWARENESS ABOUT THEM
지금부터, Las Cigarreras는 그들에게 필요한 공간을 제공할 거에요. 그리고 그들에 대한 사회적 인식도 바꿀 거에요.
CULTURE
BY NOT MONEY, BUT EVENTS!!
돈이 아닌, 그들 사이에 발생할 이벤트로 말이죠!
AMONG THEM, CHILDREN LIKE ME ARE VERY IMPORTANT! BECAUSE THEY ARE POTENTIAL COMMON PEOPLE!
이중에서도 저 같은 아이들은 중요한 유저에요! 우리는 잠재적인 일반인이잖아요!
AREN'T THE ADULT WHO HAVE GROWN IN THIS SURROUNDING DIFFERE
FROM OTHER
이런 환경에서 자라난 어른들이라면 인식이 다르지 않을까요
그리고 이 도시를 구하는 거죠!!
AND THEY'LL SAVE THIS CITY!!

aily
rogram
Through
the other
programs
Daycare
center
SO, WE DECIDED TO MAKE A DAYCARE CENTER!!! AND ARRANGED ALL THE OTHER PROGRAMS FOR USERS ON THE FLOW TO MAIN PROGRAM.
그래서 우리는 Daycare 센터를 만들기로 했어요.
그들을 위한 모든 프로그램들을 동선을 따라 배치했죠
WE REMAINED ORIGINAL LAYERS HERE, AND PUT MANY ROOMS ON THEM
기존에 있던 Las Cigarreras 층들은
그대로 남긴 채, 여러 개의 방들을 만들어 주었어요.
BUT THE PROBLEM WAS HOW TO CONNECT THEM AND MAKE CHILDREN PASS THROUGH PROGRAMS.
MARIA AND PEPA GAVE US BIG HELP!!
RAMP
하지만 아이들을 위한
각 레이어의 동선이 문제였어요.
페파와 마리아가 엄청난 도움을 주었지요!!
MY CHILD DOESN'T GO THROUGH THE PATH WHERE HE HAVE BEEN BEFORE.
AND PROF. PERNANDO ADVISED FOR US
"우리 애는 한번 갔던 길은 잘 안 간다네."
그리고 페르난도 교수님도 조언을 해주셨지요
WE TRIED PUTTING FUNNY PATHS AMONG THE RAMP
그래서 우리는 램프 사이사이 재미있는 길을 넣어 보았어요.
IT MUST BE FUN!!
AND HAD THEM PASS THROUGH THE MAIN UNITS
대표적인 유닛들을 통과하게끔요.

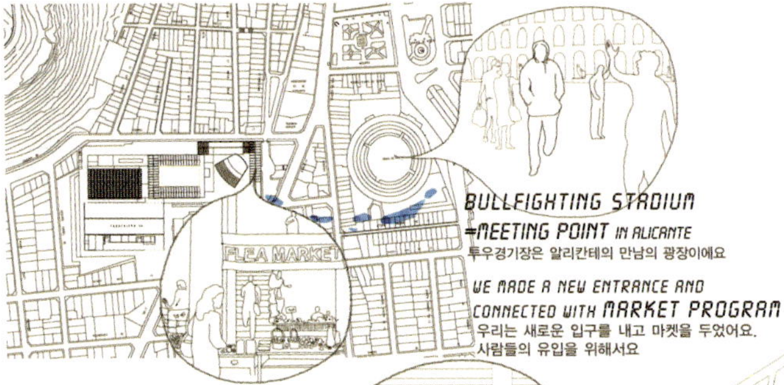
FLEA MARKET
BULLFIGHTING STADIUM
=MEETING POINT IN ALICANTE
투우경기장은 알리칸테의 만남의 광장이에요
WE MADE A NEW ENTRANCE AND
CONNECTED WITH MARKET PROGRAM
우리는 새로운 입구를 내고 마켓을 두었어요.
사람들의 유입을 위해서요

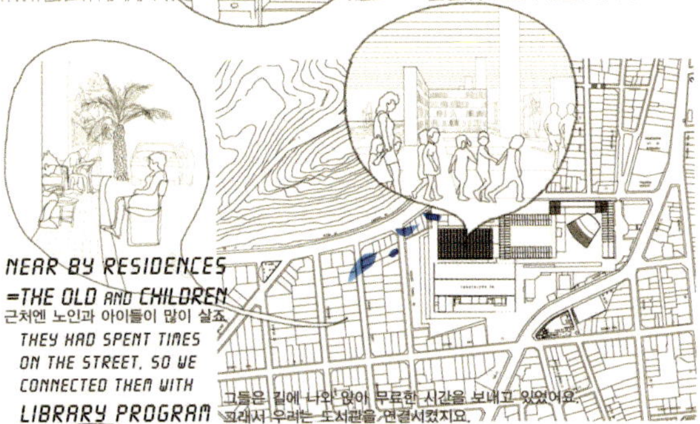
NEAR BY RESIDENCES
=THE OLD AND CHILDREN
근처엔 노인과 아이들이 많이 살죠.
THEY HAD SPENT TIMES
ON THE STREET, SO WE
CONNECTED THEM WITH
LIBRARY PROGRAM
그들은 길에 나와 앉아 무료한 시간을 보내고 있었어요.
그래서 우리는 도서관을 연결시켰지요.

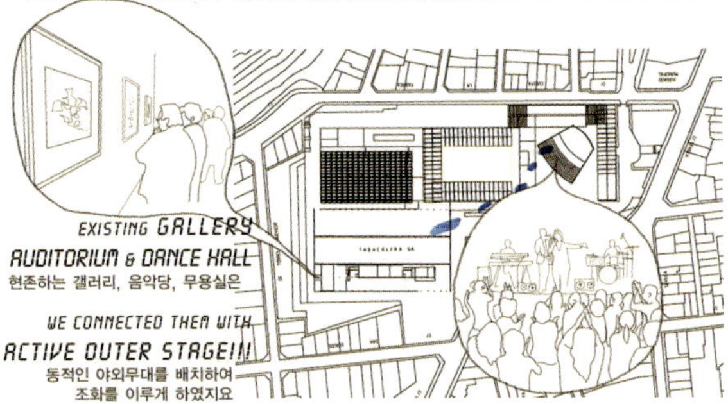
EXISTING GALLERY
AUDITORIUM & DANCE HALL
현존하는 갤러리, 음악당, 무용실은
WE CONNECTED THEM WITH
ACTIVE OUTER STAGE!!!
동적인 야외무대를 배치하여
조화를 이루게 하였지요

0
DESTINATION:
PLAY GROUND
PLZ STAND UP
AND COME WITH ME!
자, 그럼 일어나서
저랑 같이 놀러 갈까요?

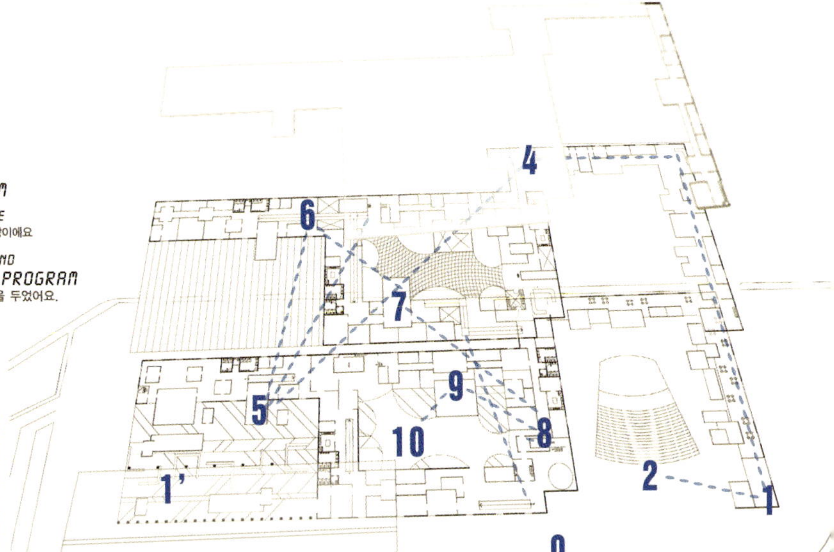
4
6
7
9
5
8
10
1'
2
1
0

OH~
GRANDMA
GO AROUND WITH
ME IN LIBRARY.
오, 할머니도 저랑
도서관가서 놀아요.

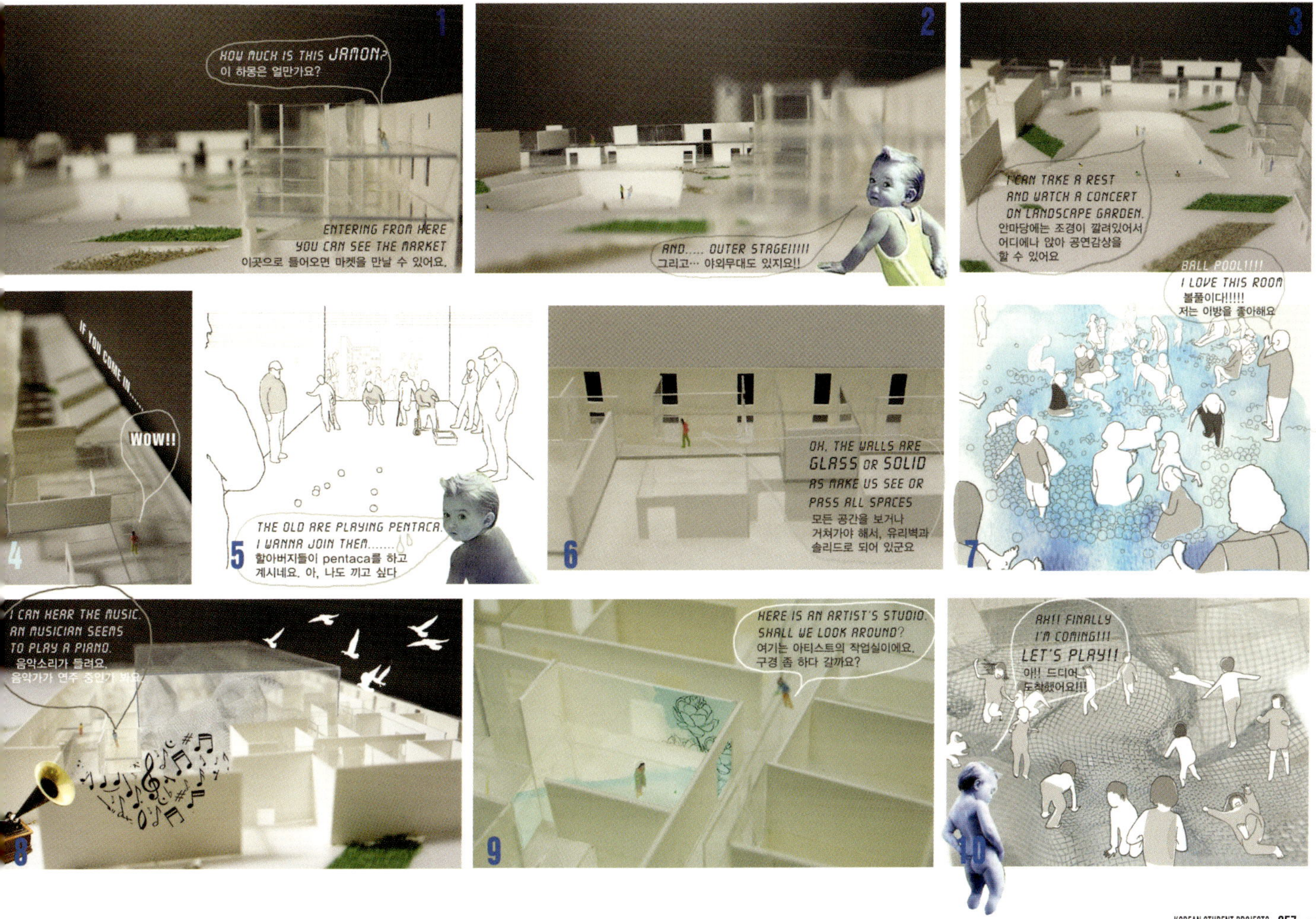
1
HOW MUCH IS THIS JAMON?
이 하몽은 얼만가요?
ENTERING FROM HERE
YOU CAN SEE THE MARKET
이곳으로 들어오면 마켓을 만날 수 있어요.
2
AND..... OUTER STAGE!!!!!
그리고··· 야외무대도 있지요!!
3
I CAN TAKE A REST
AND WATCH A CONCERT
ON LANDSCAPE GARDEN.
안마당에는 조경이 깔려있어서
어디에나 앉아 공연감상을
할 수 있어요
4
IF YOU COME IN.......
WOW!!
5
THE OLD ARE PLAYING PENTACA.
I WANNA JOIN THEM.......
할아버지들이 pentaca를 하고
계시네요. 아, 나도 끼고 싶다
6
OH, THE WALLS ARE
GLASS OR SOLID
AS MAKE US SEE OR
PASS ALL SPACES
모든 공간을 보거나
거쳐가야 해서, 유리벽과
솔리드로 되어 있군요
7
BALL POOL!!!!
I LOVE THIS ROOM
볼풀이다!!!!!
저는 이방을 좋아해요
8
I CAN HEAR THE MUSIC.
AN MUSICIAN SEEMS
TO PLAY A PIANO.
음악소리가 들려요.
9
HERE IS AN ARTIST'S STUDIO.
SHALL WE LOOK AROUND?
여기는 아티스트의 작업실이에요.
구경 좀 하다 갈까요?
10
AH!! FINALLY
I'M COMING!!!
LET'S PLAY!!
아!! 드디어
도착했어요!!!

CHANGEABLE

Let's move out!

Soyoung Kim & Eunmi Park

EXPERIENCE MAP

도시에선 지하로 다니던 tram이 지상으로 올라오자 마자 만나는 La Sangueta, 이렇듯 sangeta는 도시와 비치의 연결점에 위치하며 그 유일한 이동수단은 tram이다.

La sangueta, this is the first place when the underground tram comes out onto the land. La sangueta is located in between the city and the beach.
Besides, the tram is the only transportation in this area.

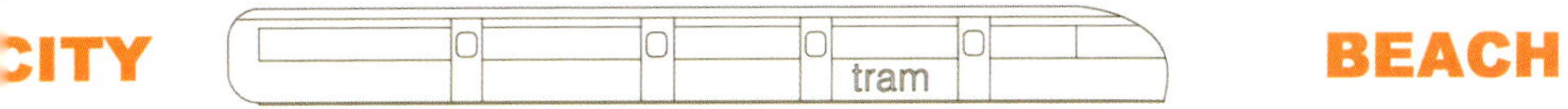

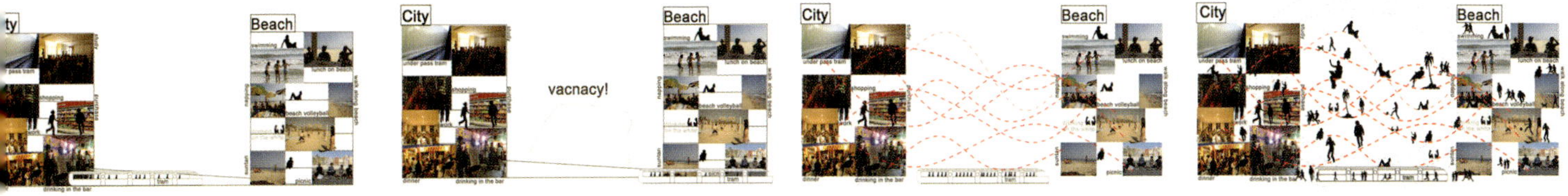

KOREAN STREET STALLS

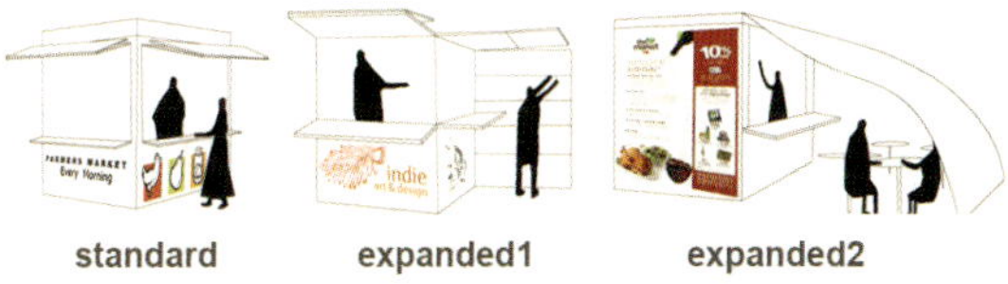

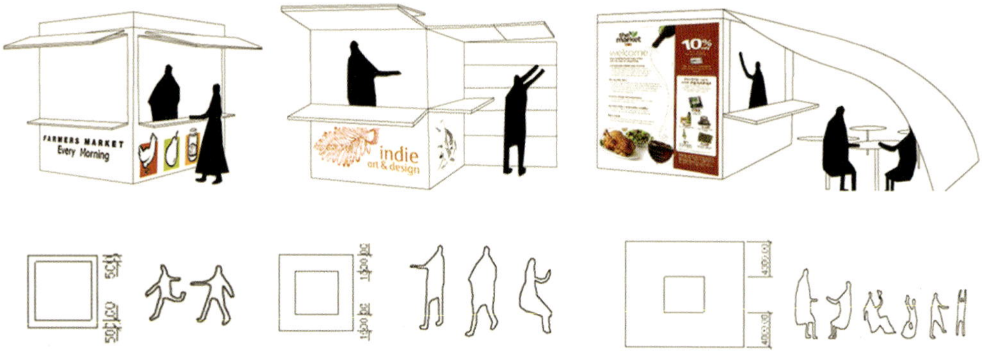

우리는 알리칸테에서 너무나도 흥미롭고 다양한 경험을 했다. 알리칸테의 해변가에서는 스페인의 뜨거운 태양을 만끽하며 해수욕을 즐기기도 하고, 알리칸테 번화가에서는 지친 심신을 풀어주는 저녁식사 또는 신나는 파티를 즐기기도 하였다.
그러나 정작 우리의 대지인 'La sangueta' 를 방문 했을 때, 우리에게 할 수 있는 것이라곤 경치를 구경하는 것 뿐이었다. 텅 빈 대지에서 느껴지는 외로움과 공허함.
우리는 이러한 대지의 현상과 느낌이 서울의 명동거리와 흡사하다고 생각했다. 명동은 자칫 여타 많은 거리들과 다를 바 없을 수도 있었을 것이다. 그러나 거리에 무수히 많은 가판대, 포장마차들로 인해 명동거리 특유의 표정과 의미를 지니게 되는 것이다.
이로써 명동에 활기를 불어넣어주는 포장마차의 특징과 La sangueta 주변의 아름다운 자연을 결합하여, 시시각각 변하고 움직이는 새로운 개념의 마켓을 제안한다.

We experienced lots of things while we were visiting Alicante. We could take a rest on the beach and work in the city. But we could not have any special activities in our site 'La sangueta'.
Since we had felt this way, we thought it might be the same situation as Myung-dong in Korea. Myung-dong is a shopping mecca in Seoul. Myung-dong is lively thanks to the Korean cart-bars. Actually the night scene in Myung-dong is what makes it really special.
We decided to combine the cart-bars in Myungdong and Alicante's beautiful environment. Finally we suggest a brand new market in Alicante which can be adapted perfectly to Alicante's situation.

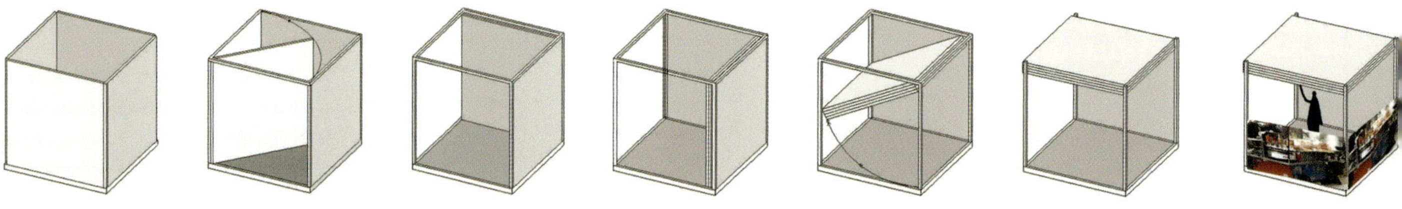

Whole Food
Nutrition
Coach

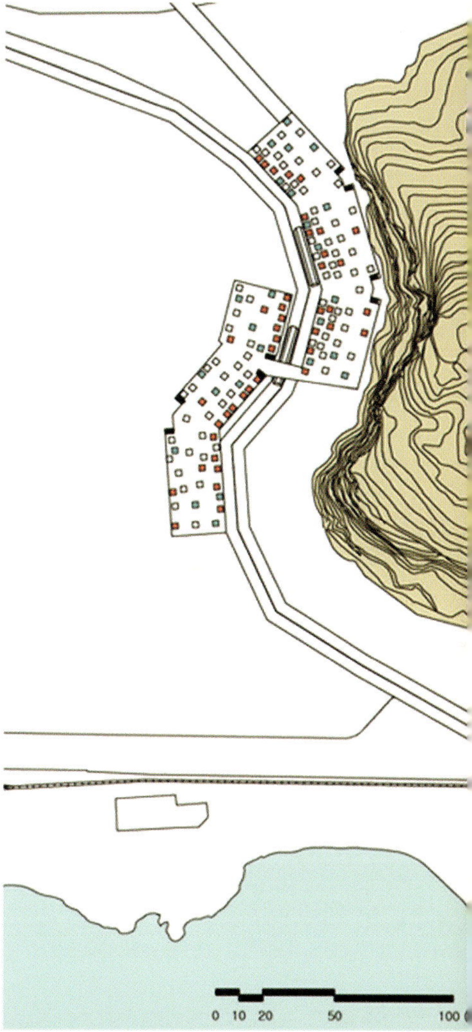
0 10 20 50 100

Open the Void

La Sangueta, Alicante, Spain

no-ryang-jin
fish market

+

harbor city
alicante

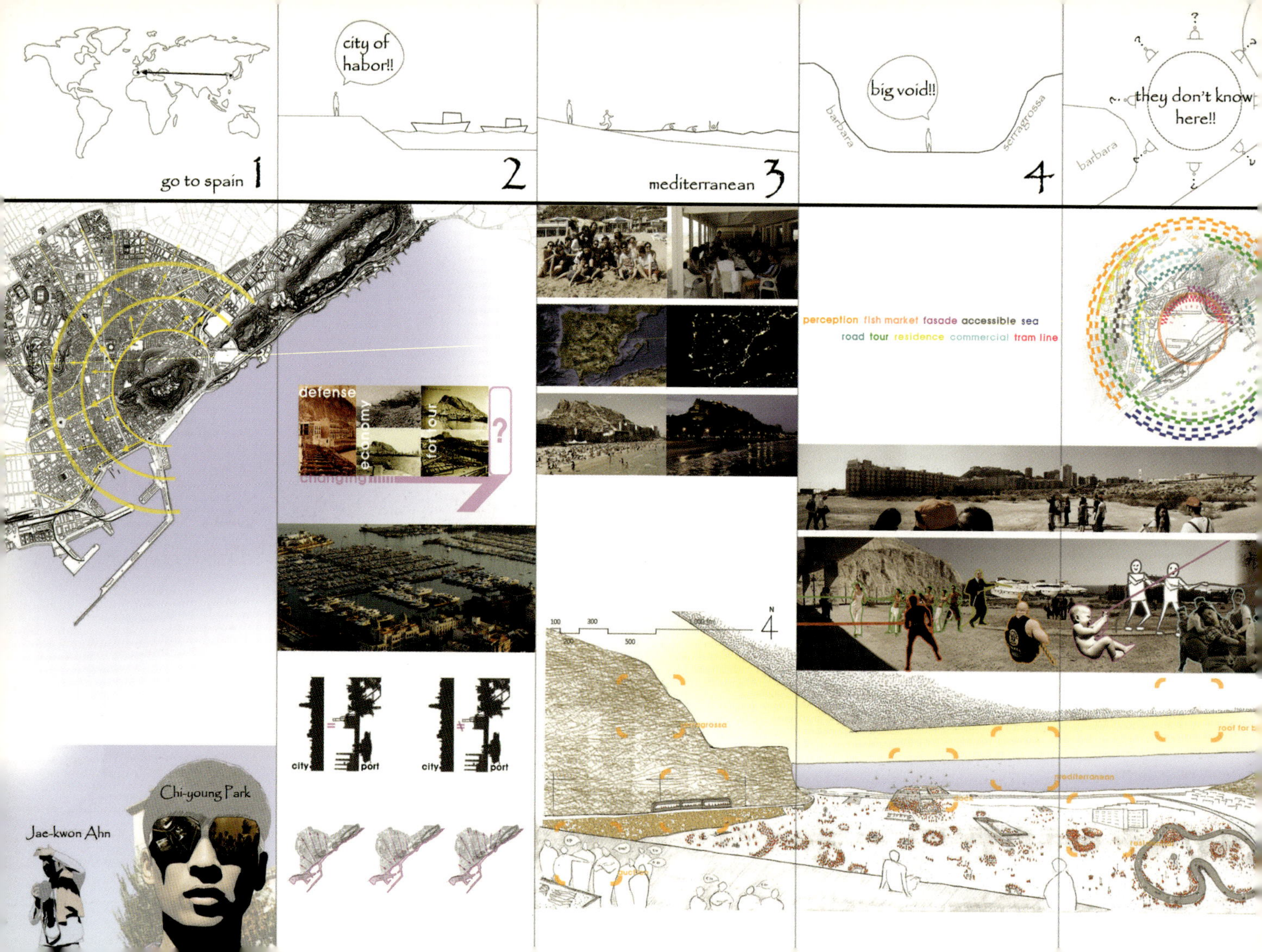
go to spain 1
city of habor!!
2
mediterranean 3
big void!!
barbara
serragrossa
4
they don't know here!!
barbara
defense
economy
tour
changing
?
perception fish market fasade accessible sea
road tour residence commercial tram line
city
port
city
port
100
300
200
500
N
4
serragrossa
mediterranean
Chi-young Park
Jae-kwon Ahn

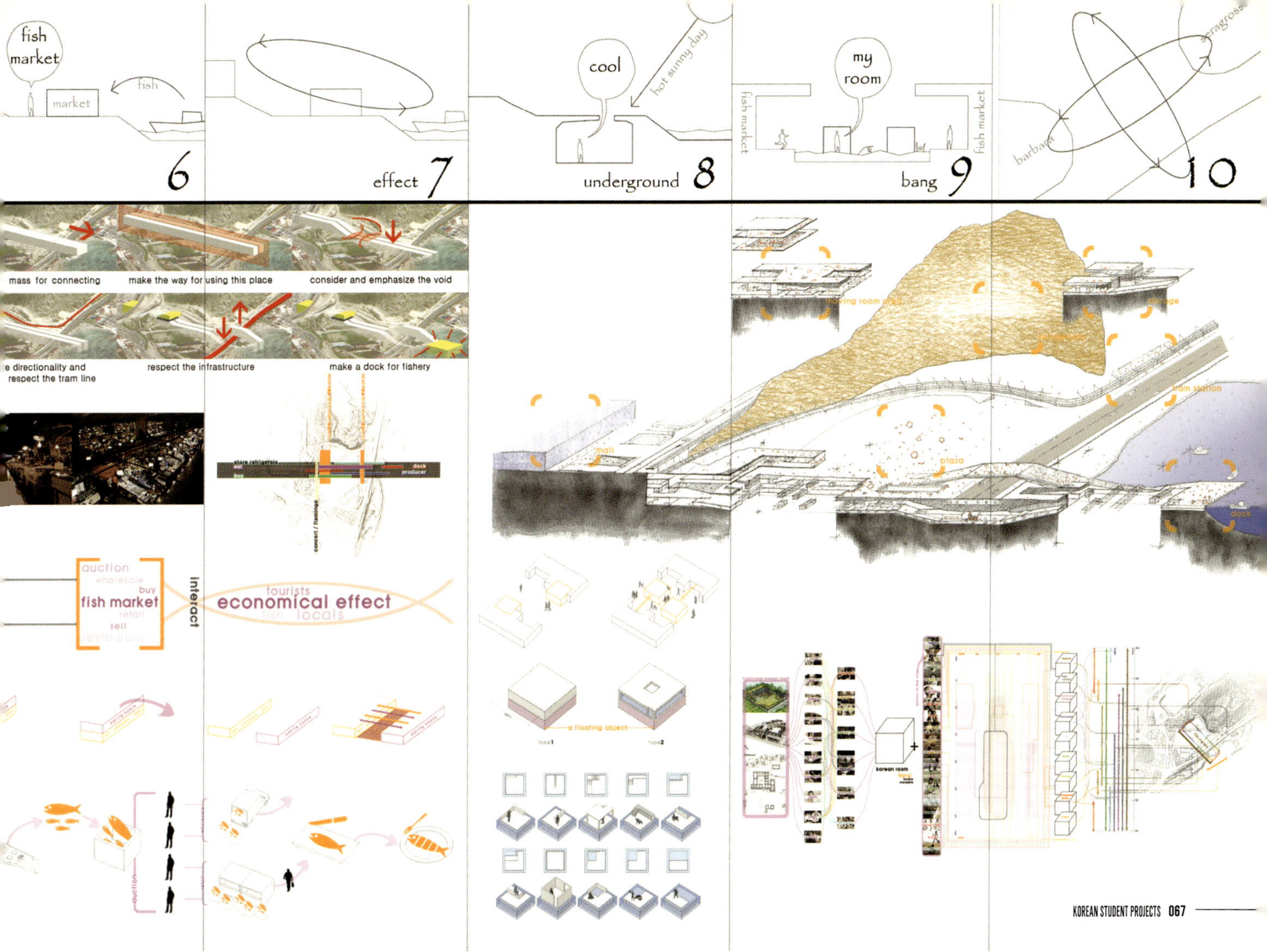

fish market
market
fish
6
effect 7
cool
hot sunny day
underground 8
my room
fish market
fish market
bang 9
barbara
10
mass for connecting
make the way for using this place
consider and emphasize the void
e directionality and respect the tram line
respect the infrastructure
make a dock for fishery
store refrigerate
eat
buy
dock
producer
concert / flamingo
mall
plaza
tram station
dock
auction
wholesale
buy
fish market
retail
sell
restaurant
interact
tourists
economical effect
fish
locals
a floating object
type1
type2
korean room
auction

FISH
FISH FISH

FISH
FISH
FISH

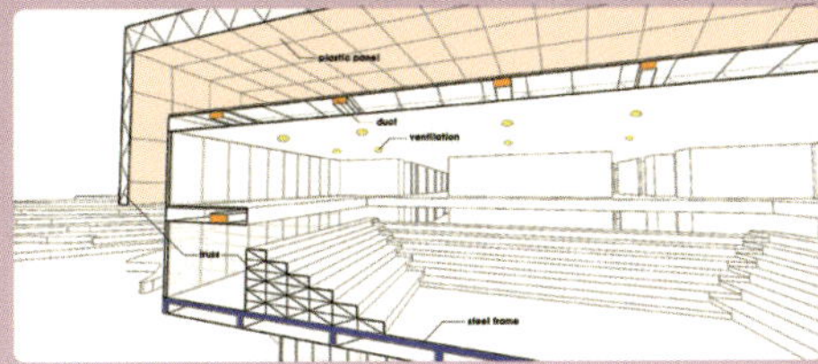
plastic panel
duct
ventilation
truss
steel frame

b1
50(m)
30
10

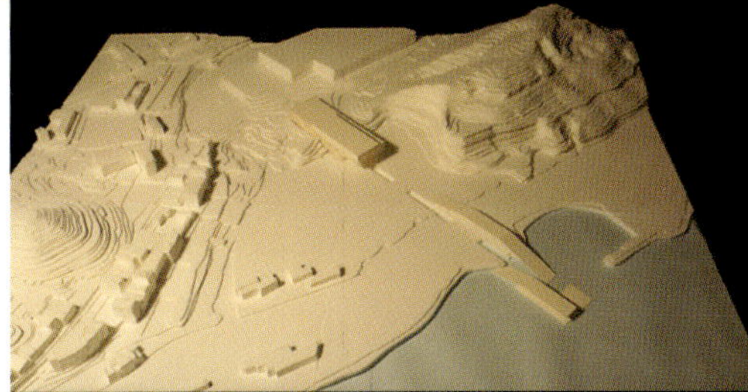

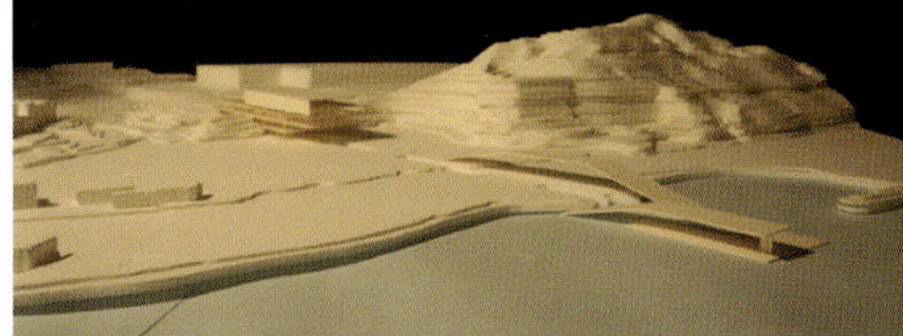

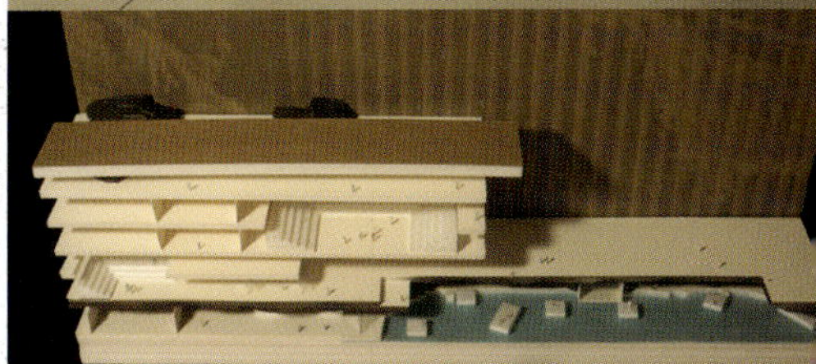

NET-WORKSHOPPING

Ivan Capdevila

As mentioned before, one of the most distinguishable bets has been that of celebrating the workshop in the street instead of the classroom.

The classroom itself is substituted by a web space for team working. In this case, the blog alicanteseoulworkshop.wordpress.com becomes the meeting point where the students share both experiences prior the visits and feedback for later work developments. Besides, this is complemented by the creation of several Facebook groups where different kind of relationships between the collective take place.

Therefore, our proposal is to replace the format of short duration workshop - concentrated on a few days experience - by a format of a long duration workshop - four months. Those days of living together are used to establish affective and emotional links not only between the students themselves but these ones and their sites. We must remember the sites will not be seen as backdrops but places where 'little extraordinary stories' happen.

All these multiple emotional links are a guarantee for the success of this format, given that a certain medium intensity but continuous dedication is required.

Alejandro Chichizola Olivieri 2011년 11월 5일

Hi guys!!
I've red your blog with your work and I think it's really amazing all the development. If you guys want to follow the way of botellon, it's right, but I miss some stuff...
I love the image of "interpretation of botellon" but you have missed all the other pedestrian that you have been developing
I would love to see a really really huge map of Alican
connections you've done with the botellon but with a
together in the same map. Otherwise, if you guys or
site to make botellon, I'm quitly sure it will not be use
people who usually make botellon, because to make
any simple street.
Maybe the question is: why I will go there to do a bo
right here (in a closer place) ??
Maybe the interaction between the diferent time anc
people make people go there and not to the side of t
On the other hand: do not forget the map youive al
connection of the botellon because I think that the n
site is the connection with the city!!!

please, ask me whatever you want!!

Hope I've been useful!!
good luck with your work!

Belen Gomez Llobell
Hey guys! We are searching a name for our
connect it with the Korean workshop. As we
word that sounds like this, or any other exp
hamar. We are open to any suggestions=)!
좋아요 · 댓글 달기 · 게시물 받아보기 ·

Mary Pepa Ramos Mira, Gala Calatayu
아합니다.

Gala Calatayud help us chiyoun
2011년 9월 29일 오전 3:13 · 좋

Hyemi Park What is hamar? Haha. In Korea [hamar] means a hippo:-)
2011년 9월 29일 오전 8:48 · 좋아요 · 1

Adrian Beltran our project group is call "hamarkun"
2011년 9월 29일 오전 9:46 · 좋아요

박지현
Hi. my friends. Now we need your helps. We have to start making contour models of two site where we've been expolring in your city, La Sangueta & Las Cigarreras. Now we guys have some difficuties to begin our these works because of lack of information. We got some CADD files from your professor indicating your city map. However It is too simple to make its contour models. So I want you guys to send some detail files & information. I heard that you guys had finished some projects on these sites. Could you guys help us? Please ~~ I love you. Thank you.~
좋아요 · 댓글 달기 · 게시물 받아보기 · 2011년 10월 13일 오후 4:56

Alejandro Chichizola Olivieri I've never worked in those sites but I have a CAD file of Alicante. Maybe it could help you guys!! I do not know what Ivan gave to you.
Tell me if you need something "better"...

kisses!!
2011년 10월 14일 오전 5:27 · 좋아요 취소 · 1

Alejandro Chichizola Olivieri I do not know if it is the map that Ivan gave to you...but...just try!!
link>>> http://www.megaupload.com/?d=1CDAM9SU

MEGAUPLOAD - The leading online storage and file delivery service
www.megaupload.com

2011년 10월 14일 오전 5:40 · 좋아요 취소 · 2

Javier Esquiva López Hi guys, I have a 3d model to the Cigarreras.
2011년 10월 15일 오전 5:52 모바일에서 · 좋아요 ·

Adrian Beltran http://www.alicante-ayto.es/redir.ph
apartado=urbanismo&pagina=plangeneralprov2%
2Fhome.html&titulo=Plan+General+%28Exposici%F3n
blica%29+-+Urbanismo+%2F+Ayuntamiento+de+Alic
2011년 10월 15일 오전 7:27 · 좋아요 · 1

Adrian Beltran hi guys, in this web site you can find
some maps of the city,just click on "planos de ordenaci
pormenorizada" and using this programme: "aide PDF t
converte" you guys can change from pdf you CADD file
easy.
2011년 10월 15일 오전 7:32 · 좋아요

박지현 Thank you Adrian!
2011년 10월 15일 오후 1:14 · 좋아요

Mary Pepa Ramos Mira
hey guys!!! how r u doing?
i need a favour, we need t
anyone of u send them to r
them, post it? we would re
thanxxx 1000 kisses
좋아요 · 댓글 달기 · 게시

ByungYup Lee님이 좋

Hyemi Park I ca
are in. Could u let
2011년 10월 20

Mary Pepa Ran
marypepa_pinoso
2011년 10월 20

Soyoung Kim
just 3 days left! we r waiting for u guys with huge expectation :D plz be well pre
for coldness!! — 박지현님 외 11명과 함께

댓글 달기 · 게시물 그만받기 · 2011년 11월 2일 오후 8:52

Gi Es님과 Minjae Kim님이 좋아합니다.

입력하세요...

...no se ha apuntado anteriormente, una de las apuestas más diferenciales es la de celebrar el taller ...a calle y no en el aula.

...aula se sustituye por un espacio web de trabajo en equipo. En este caso, el blog ...anteseoulworkshop.wordpress.com se convierte en el lugar de encuentro en el que los ...diantes comparten tanto experiencias previas a las visitas como aportaciones al desarrollo de los ...pectivos trabajos. Además, esto se complementa con la creación de varios grupos de Facebook ...os que se producen distintos niveles de relaciones entre el colectivo implicado.

...la propuesta es la de sustituir el formato de taller de corta duración - concentrado en una ...eriencia de días- por un formato de taller de larga duración - cuatro meses- en los que los días de ...vivencia se usan para establecer lazos afectivos y emocionales no sólo entre los propios ...diantes sino entre los estudiantes y los lugares de trabajo, ya que estos, recordemos, no serán ...os sólo como escenarios sino como lugares en los que tienen lugar "pequeñas historias ...aordinarias".

...stablecimiento de estos múltiples lazos emocionales es una garantía de éxito para este formato, ...l que se requiere cierta dedicación de intensidad media pero diferida en el tiempo.

can attract people not the boats or ships. After, we had reminded our experiences in Spain(Alicante) such as night culture, early closing market and having meal for a very long time.

Based on these two processes, we are going to suggest 24hours opened fishery marcado(market) as a cultural harbor. The important thing is the place is going to change as time goes. It can be a market, restaurant and aquarium at daytime and pubs(club), a huge hall(for the concert or dance party) and pool at night time. Therefore, the place is going to activate for 24hours with various of people(different ages).

Well, we are still working on the project but I want to know your opinions as a native Spanish. Please give us some comments if there are something we are missing. Thank you guys. I love you so much.

좋아요 · 댓글 달기 · 게시물 그만받기 · 2011년 11월 2일 오전 3:33 서울특별시 근처에서

4개의 댓글 모두 보기

Ivan Capdevila Vicente Iborra About the proposed use I would be very careful. Note we have plenty of leisure along the harbor. We also have a marina, and many uses that provide almost a 24H use. Note this leisure area is located exactly where the city center meets the s...

더 보기

2011년 11월 2일 오후 9:06 · 좋아요

박치영 Thank you for your comments, Ivan.

2011년 11월 2일 오후 9:11 · 좋아요

댓글을 입력하세요...

Sara Gi...

Hi Seong... It looks ... these da... one with... watching... hahahha... awesom... very acc... it! Btw, ... building. Sometim... route of... where th... people g... another ... worst! b... love to g... to avoid... What ar...

...ot wait to go!! ...korean apartment, could ...newhere where we can find

...월 20일 오전 2:34

...that apartment plans ...drass? I will send u!! 좋아요

...eee!!! love ya

...nd u it right now!

:-) 좋아요

...eeee!!!! :D

앞서 언급한 바와 같이, 가장 구별되는 도전은 워크샵이 교실에서가 아니라, 도시의 거리에서 행해지고 있는 것이다.

교실의 공간 자체는 팀워크를 위한 웹스페이스에 의해 대용되는 것이다. 이 경우에는, 블로그인 'alicanteseoulworkshop.wordpress.com'는 학생들이 이전의 서로의 국가를 방문하고 이후 진행하는 작업의 피드백을 공유하는 만남의 장소가 되었다. 게다가, 몇몇의 소그룹이 페이스북에서 형성되어 블로그에서 모자란 점들을 보충해 주고 있다.

이런 이유로, 단지 몇 일간의 고도로 집중된 짧은 기간 동안의 형식이 4달 정도의 장기 워크샵으로 대체되어야 한다는 것이 우리의 제안이다.

함께 생활한 시간들은 학생들 사이의 정서적 교감뿐만 아니라 주어진 대지에 대한 정서적, 감정적 교류를 성립하였다. 우리는 이 대지가 단지 배경으로 작용하는 게 아니라 작지만 특별한 이야기가 만들어지는 장소로 보여 질 수 있다는 것을 상기해야 한다.

여기에 모든 복합적인 감성의 연결들이 어떤 매개체의 집중과 헌신으로 지속된다면 이러한 형식의 성공을 보증하고 있다.

'12 unforgettable cities within Seoul'

by Gala Calatayud, Adrián Beltrán, Mary Pepa Ramos and Belén Gómez

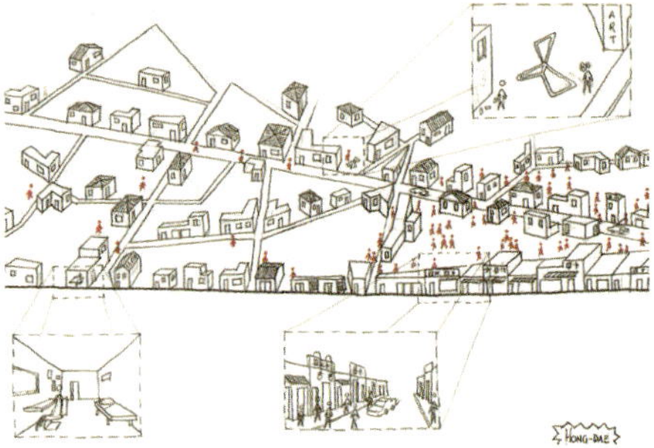

1. Hong Dae

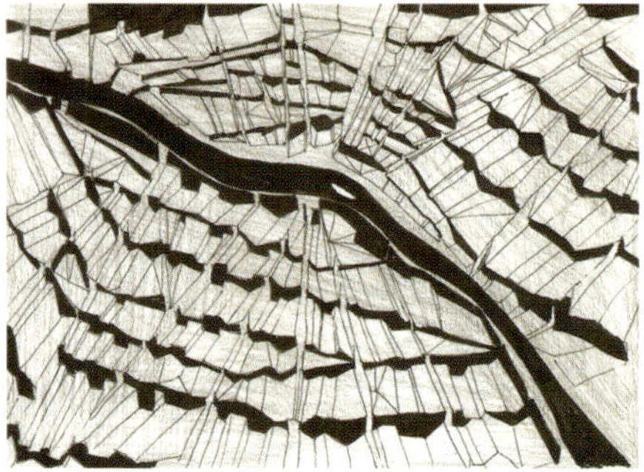

2. Nodle Island

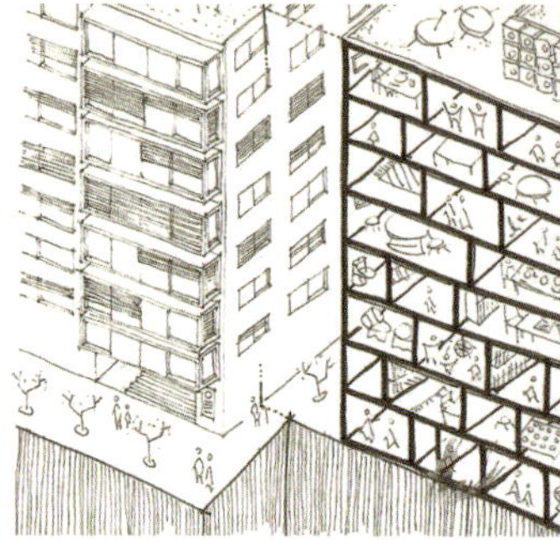

3. Apartment Complex in Ichon-dong

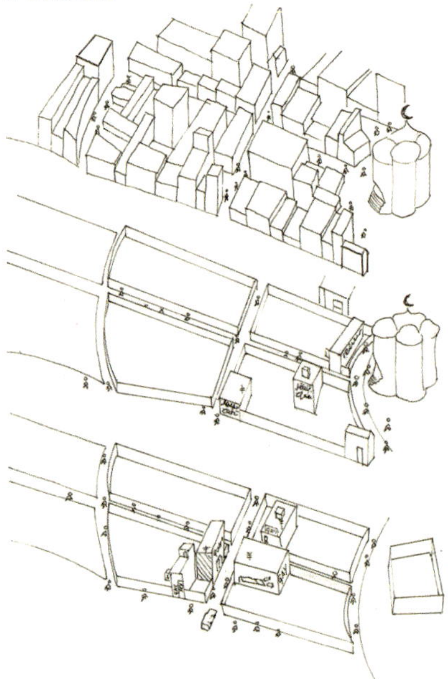

4. Itaewon

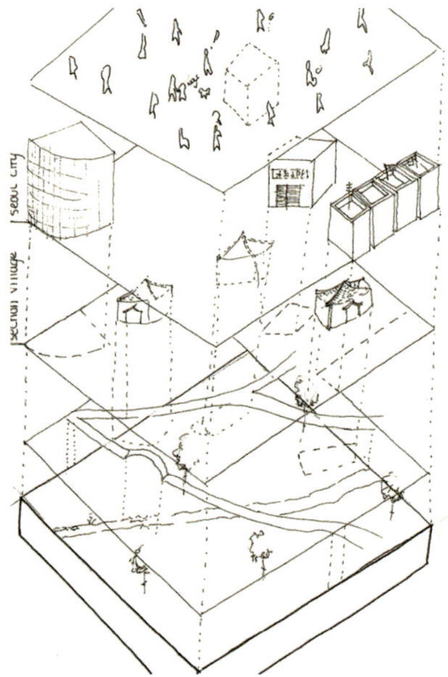

5. Seochon Village

6. Dongdaemun Market

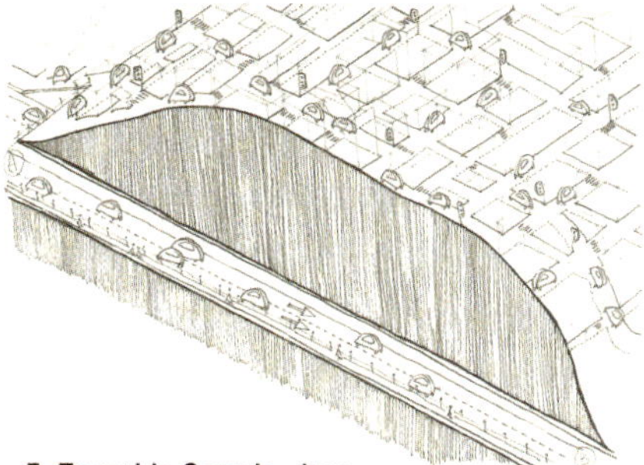

7. Tunnel in Sangdo-dong

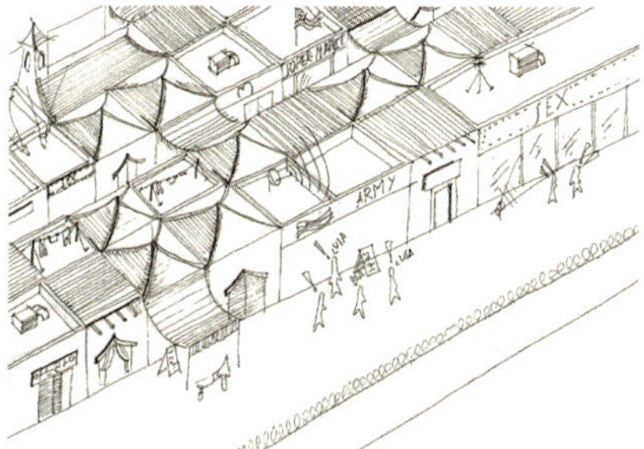

8. Haebang chon

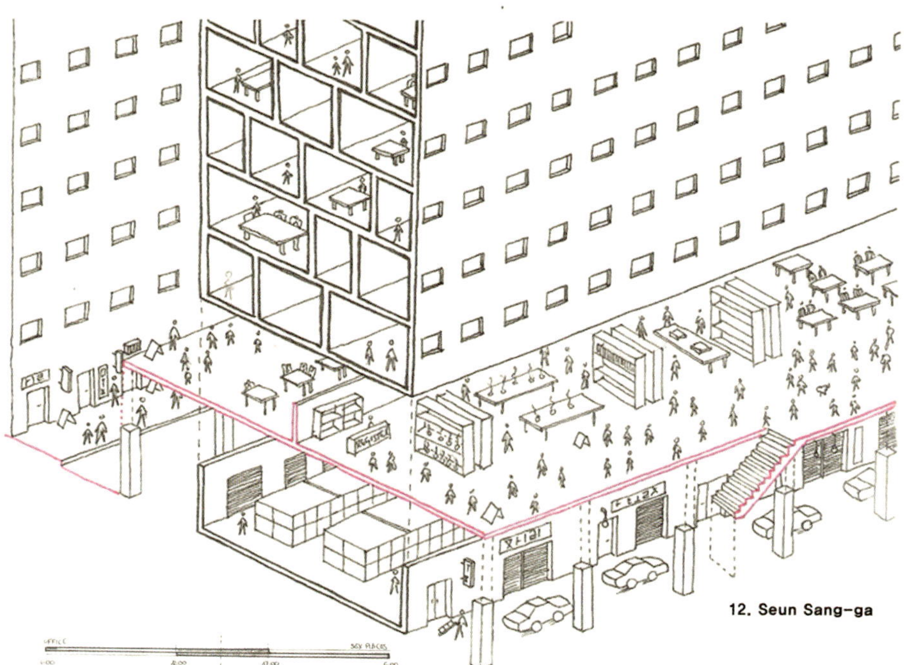

12. Seun Sang-ga

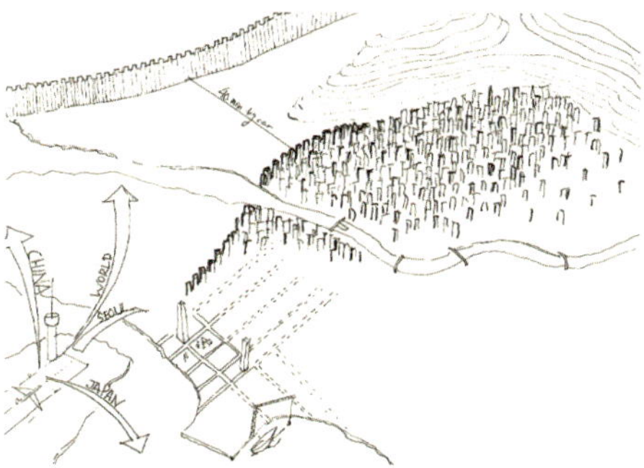

9. Song-do New City

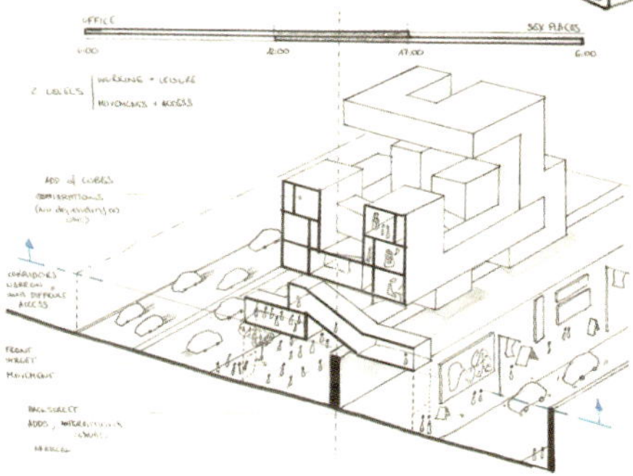

10. Teheranno

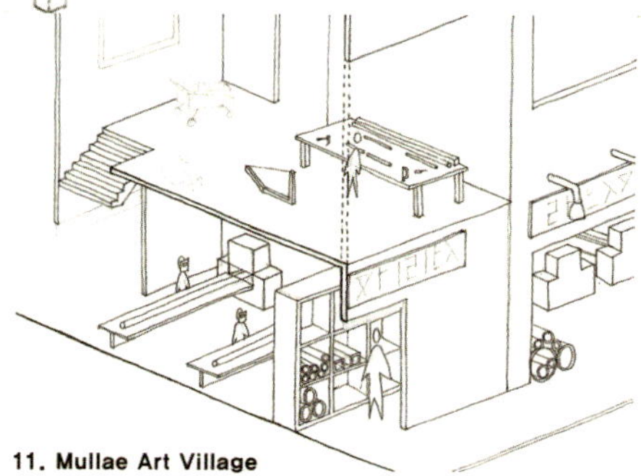

11. Mullae Art Village

Seoul

서울은 대한민국의 수도이며 가장 큰 도시이다.
대한민국 절반이상의 인구가 서울반경안에 살고 있고
국가의 경제,정치, 문화의 주를 만들고 있는 상황이다.

찜질방, 씻고 먹고 즐기고 잠자기.
The Jimjilbang, Wash! Eat! Play! and Sleep!

놀랍게도 노래방이라는 곳의 안과 밖에서, 한국 학생들은 마치 다른 사람 같았다.
Surprisingly, Korean students were so different in Noraebang.

苦盡甘來, 고생끝에 낙이온다. 처음은 쓰게 끝은 달게!
Pain is gone, and pleasure is come. Sweet after bitter!

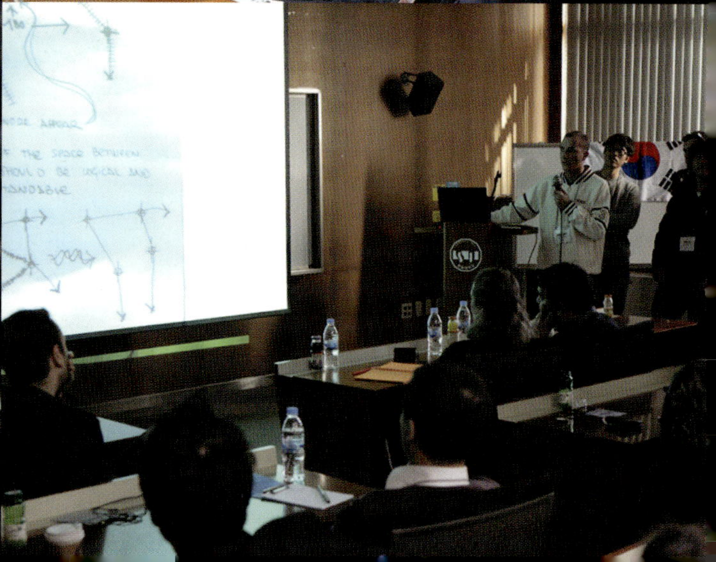

2011 Soongsil - Alicante Workshop

SOONGSIL ARCHITECTURE THURSDAY LECTURE
2011-2
5TH
"Learning from the Urban"
2011.12.08
5:30-7:30 PM
벤처관 309호
Iván Capdevila
Founding partner of PLAYstudio
www.playstudio.es

숭실대학교

SEOUL: DIARY OF A NON RETURN TRIP

Ivan Capdevila

Jungin Kim's proposal of working about the "bang" concept considerably conditions the later events.

We must remember "bang" is the Korean word for describing the domestic space. The language evolution has created name of places where the domestic space has been stretched into the public-collective one. Therefore, places as "Noraebang" - the Korean karaoke- or "Jimjilbang" - a sort of collective sleeping room- are places where the domestic is collectively celebrated.

Such intense relationship between the domestic and collective, that the language itself evidences, has been one of the most determining aspects in the detailed approach by the students.

While Gala shows us how the "Noraebang" building configures its urban facade like a projection towards the exterior of the different domestic-scaled and "hyper decorated" karaokes, Maria and Eduardo show us an extremely complex and self-sufficient inner world, the "Jimjilbang" : a place where nudity and corporal purification is a previous step to socialization. The pyjama is the second relational device. It removes any social class or categorical feature. This is necessary for enabling a horizontal interaction, something completely new in a society where every manifestation is hierarchized. Maria's Jimjilbang drawing shows us perfectly this is the place where domestic situations like watching TV, eating, speaking, playing or, even, sleeping are celebrated collectively.

The common characteristic of both places is a very sophisticated and broad use of technology for their functioning. In both cases, technology is necessary as an instrument for socialization.

The Korean restaurant is other of these spaces of social interaction. Its design as space of comfort (and home's evolution) will be a subject of detailed study by Cristina. The step, heated floor, tables as floor extrusions, walls partitions, use of mobile dividers and so on will be devices analysed by this drawing in order to understand the construction of this domestic comfort.

Other of the experienced architectural devices in the construction of social relations is soyu, alcoholic drink, but more in particular, its bottle. Maria studies how serving soyu around the table is affected by hierarchical conditions -defined by the age of those sharing the table- and how this affects the later events. Instead, Alejandro analyses the bottle as a game mechanism and whose cap -specially its opening system and number written at the back- enables a series of horizontal relations between those around the table.

Both cases, analyse the architectural ability of soyu, both as a serving protocol and game mechanism, of generating socializing situations within a context designed as a highly domestic space.

Cristina, on the other hand, shows us how the restaurant table, understood as a minimum and repeatable module, is perfectly designed to determine the maximum number of people according to their use of dishes, glasses, and so on.

Hugo, Javier and Adrian get us closer, at the end, to a personal domestic space experienced in a direct and different way.

Hugo, on the one hand, studies how the architecture determines the relations between him, his host and his host's grandmother. On the other hand, Hugo unveils how the wardrobes location in public places of the house ends up modifying his own wardrobe to the extent he becomes his host.

Javier describes an intensive use of space and how the objects invasion conditions it.

Adrian, unlike Javier, shows us a versatile use, understanding the multiple spatial relations simultaneously happening in a single space.

Beyond this point, Cristina presents us the underground wagon as another home. Due to the everyday long stay inside means of transport, citizens include this precious time within their domestic routines. Inside the wagon, one can keep on sleeping, work or start buying. This is configured, thus, as an intermediary space in-between home and the city.

In parallel, Sara steps back and studies which characteristic aspects of the traditional house are present in those contemporary places their companions are drawing.

Sara also opens the door to other places of social interaction. She presents the Seoul Tower as a mechanism of intermediation between couples or, in her own words, as a mechanism for love reinforcement. Her reading has nothing to do with the iconographic but with its emotional use as well as how an architecture of this kind may end up being used as a place where love is sold and celebrated collectively.

Other commercial places are Eduardo, Belen and Mary Pepa's ones.

Eduardo draws a street of multiple uses, scales, sizes, situations; this is a street where everything can happen.

Belen shows a commercial street typology that is reversible. During the day, it is used as a retail outlet;during the night, as a wholesale.

Mary Pepa returns to the domestic space. She shows us a commercial street configured by houses taking out their objects to the outside. In other of her drawings, the consumption objects themselves and their uncontrolled accumulation configure the inner landscape in shopping centres.

Finally, Belen goes on a graphical journey to North Korea. This poetical approach shows us the different architectures building up an emotional landscape, by detailing the architectural mechanisms that hide the war evidences behind a musical score.

In the end, all the following drawings strive to find out the relations between situations, the culture where they take place and the architectural devices that enable them.

서울 : 귀로가 없는 여정의 일기

이반 카프데비아

김정인 교수의 '방' 개념에 대한 제안은 이후 많은 사건을 좌우하였다.

'방' 이라는 단어는 한국의 개인적인 실내 공간을 설명하기 위한 것이라는 것을 알아야만 한다. 이 단어의 발전은 사적인 공간이 다수의 공공적인 공간으로 확장되어 가면서 새로운 장소의 이름을 만들어 왔다고 생각된다. 그러므로 노래방(한국식 가라오케), 찜질방(어떤 공동 취침공간)같은 공간들은 집합적으로 사적인 행위가 허용되는 공공의 장소이다.

이런 개인적 공간과 공공 공간의 집약적 관계인 '방' 은 단어가 의미하듯이 학생들의 섬세한 접근이 필요한 가장 확실한 (경험)요소 중의 하나가 된다.

갈라는 어떻게 노래방의 건축형태가 스케일이 다른 외부의 상업공간과 주변의 현란하게 장식된 가라오케들로 투사하듯이 보이는 도시적 입면을 형성하는지 보여준 반면, 마리아와 에드와르도는 극도로 복잡하고 자급자족적인 내부세계인, 사회화 이전 단계의 장소로서 맨몸의 노출과 육체적 정화의 공간인 찜질방을 보여준다. 여기서 '파자마' 는 주목할 만한 요소이다. 이것을 입는 것은 어떤 사회적 계급이나 소속을 제거하는 것으로 수평적인 관계를 가능하게 하고 거의 모든 것들이 위계화된 사회에서 뭔가 전혀 새로운 어떤 것으로 보인다. Maria의 찜질방 드로잉은 TV시청, 식사, 대화, 게임 심지어 수면까지 사적인 상황이 단체적으로 이루어지는 공간임을 완벽하게 보여준다.

두 공간의 보편적 특징은 굉장히 섬세하고 광범위 한 테크놀로지의 이용이 이 장소들의 기능을 충족시키기 위해 존재한다는 것이다. 두 경우 다 테크놀로지는 사회화를 위한 도구로서 필수적이다.

한국의 식당은 사회적 상호작용 공간의 다른 예이다. 가정(집)에서 출발 한 것 같은 휴식의 공간으로서 이 장소는 '크리스티나'에 의해 세밀히 연구될 주제이다. 계단, 온돌, 층의 연장으로서 테이블, 벽 파티션, 이동식 칸막이의 사용 등을 드로잉을 통해 살펴봄으로서 가정과 같은 휴식의 구조를 이해하게 될 것이다.

주류인 소주, 구체적으로 소주병은 또 다르게 사회적 관계를 형성하는 경험의 건축적 장치의 한 형태다. 마리아 는 어떻게 한 테이블을 공유하는 사람들 사이에 연령에 의해 규정된 위계적 신분에 소주가 작용하는지, 그리고 어떻게 이후의 사건들에게 영향을 미치는지 연구하였다. 반면에 알렉한드로는 소주병과 그 뚜껑 뒷면에 드러난 숫자놀이를 통하여 여러 양태의 수평적 관계가 테이블에 앉은 사람들 사이에서 어떻게 가능한지 분석하였다.

소주의 건축적 능력을 분석하는 두 가지 경우, 모두 매우 개인적이며 친근하게 디자인된 공간에서 사회화를 진행시키는 상황기제로 상대를 대하는 규범과 놀이의 형태를 분석하고 있다. 한편 크리스티나는 최소의 면적과 반복되는 모듈로서 인식되는 식당테이블이 접시, 물잔 등의 용도를 맞추며, 어떻게 최대의 인원을 완벽히 수용하는 디자인이 되는지 보여준다.

마지막으로 우고, 하비에르, 아드리안은 직접적이고 다양한 방법으로 경험한 개인적이고 사적인 공간을 우리에게 좀 더 가까이 드러내 보여준다. 휴고는 어떻게 건축적 장치가 그와 그의 호스트, 그리고 호스트의 할머니의 관계를 결정할수 있는지 연구하였다. 여기서 휴고는 어떻게 가정의 공적인 공간에 놓인 간이옷장의 위치가 결국 그 자신의 옷장의 위치를 변화시켜 그가 손님이 될 수 있도록 하는지 보여준다. 하비에르는 공간의 복합적인 사용과 어떻게 이러한 다양한 사물의 수용이 전체공간을 변화시킬 수 있는지 기술하였다

하비에르와는 달리 아드리안은 하나의 공간에서 동시에 일어나는 다양한 관계를 이해함으로서 공간이 유동적으로 적응될 수 있는 예을 보여준다.

더욱이, 크리스티나는 또 하나의 집으로서의 지하철 공간을 보여준다. 매일 긴 시간 동안 대중교통 속에서 지내야 하기 때문에, 그 귀중한 시간을 이용하여 도시민들은 사적으로 행해지는 일상적 행위를 이 공간안에 포함하고 있다. 그래서 지하철은 집과 도시 사이의 매개 공간으로서 표현되었다.

더불어 사라는 어떤 전통가옥의 특징이 이런 현대적 장소에서 나타나는지 약간의 거리를 가진 관찰자의 시점으로 연구하였다.

사라는 또한 이런 현대적 장소들의 사회적 상호작용에 다른 가능성을 보여주었다. 그녀는 남산타워를 연인사이의 사랑의 강화를 위한 연결 장치로 표현하였다. 그녀의 건물에 대한 분석은 건축의 상징적 의미와는 전혀 연관이 없다, 대신 정서적인 사용과 어떻게 이런 종류의 건축이 여러 사람들에 의해 사랑을 공고히 하고 기억하는 공간으로 쓰이는지를 연계하였다.

다른 상업공간에 대한 분석은 에두아르도, 벨렌,과 마리페파에 의해 이루어 졌다.

에두아드도는 거리의 용도, 스케일, 크키, 상황 등 다중적인 모습을 그렸는데 이는 한국의 거리가 모든 것들을 담을 수 있는 곳임을 의미한다.

벨렌은 다시 원래 상태로 돌아갈 수 있는 상업지구의 유형을 보여주었다. 낮동안 동대문은 소규모 아울렛으로, 저녁에는 도매시장으로서 이용된다.

마리페파는 개인적이고 사적인 공간으로 회귀했다. 그녀는 집밖에 내어놓은 잡동사니에 의해 형성된 상업 거리를 보여주었다. 그녀의 다른 그림에서는, 소비재 자체와 그것들의 통제될 수 없는 축적상태가 쇼핑센터 안의 내부 환경을 (랜드스케이프) 형성함을 보여준다.

끝으로 벨렌은 북한접경 지역으로 시각적인 여정을 계속하였다. 그녀의 시적인 접근은 오선지 뒤에 숨겨져 전쟁의 흔적을 감추고 있는 건축적 메커니즘을 섬세한 시선으로 묘사함으로서 여정동안 여러 다른 건축물이 감성적 랜드스케이프를 형성함을 보여주었다.

결론적으로, 모든 드로잉은 사건들이 일어나는 상황들과 그것을 담는 문화 그리고 그것들을 가능하게 하는 건축적 도구들 사이의 관계를 알아내기 위한 노력의 결과물이다.

Namsan Seoul Tower.

A STORY MADE FOR LOVERS

Sara Gimenez Espejo

I remember how difficult it was to reach the top of Seoul Tower. It really was like going to visit the stars. What most surprised us was the romance of the path of architectural devices. A classic way of promoting the love between lovers. First, we had a technology room for taking pictures with the tower at night. Later, while waiting for the elevator, you get a wonderful film about the fantastic views of the tower. When time had elapsed, you're ready to take the elevator headed for the moon with your partner. Within the same elevator the receptionist will show you the path to the get to the moon. All times are unforgettable and unique. Once arrived, you're ready to see the entire 360 degree Seoul with the telescope. After the experience you can give to your girlfriend hygiene things and take your photos with Korean traditional masks. In the shops, are also offered hearts that are attached with lock to bind your love forever with your partner. With tiles decorated by the love you show your love to the world. Finally we find the sloping benches. Benches are made to have an slope. The intention is to be the closest to your partner. At the end of the route you could take a picture with the giant heart's figure and go home happily after having regained your girl a little more. What's nicer than taking your girl to see the stars?

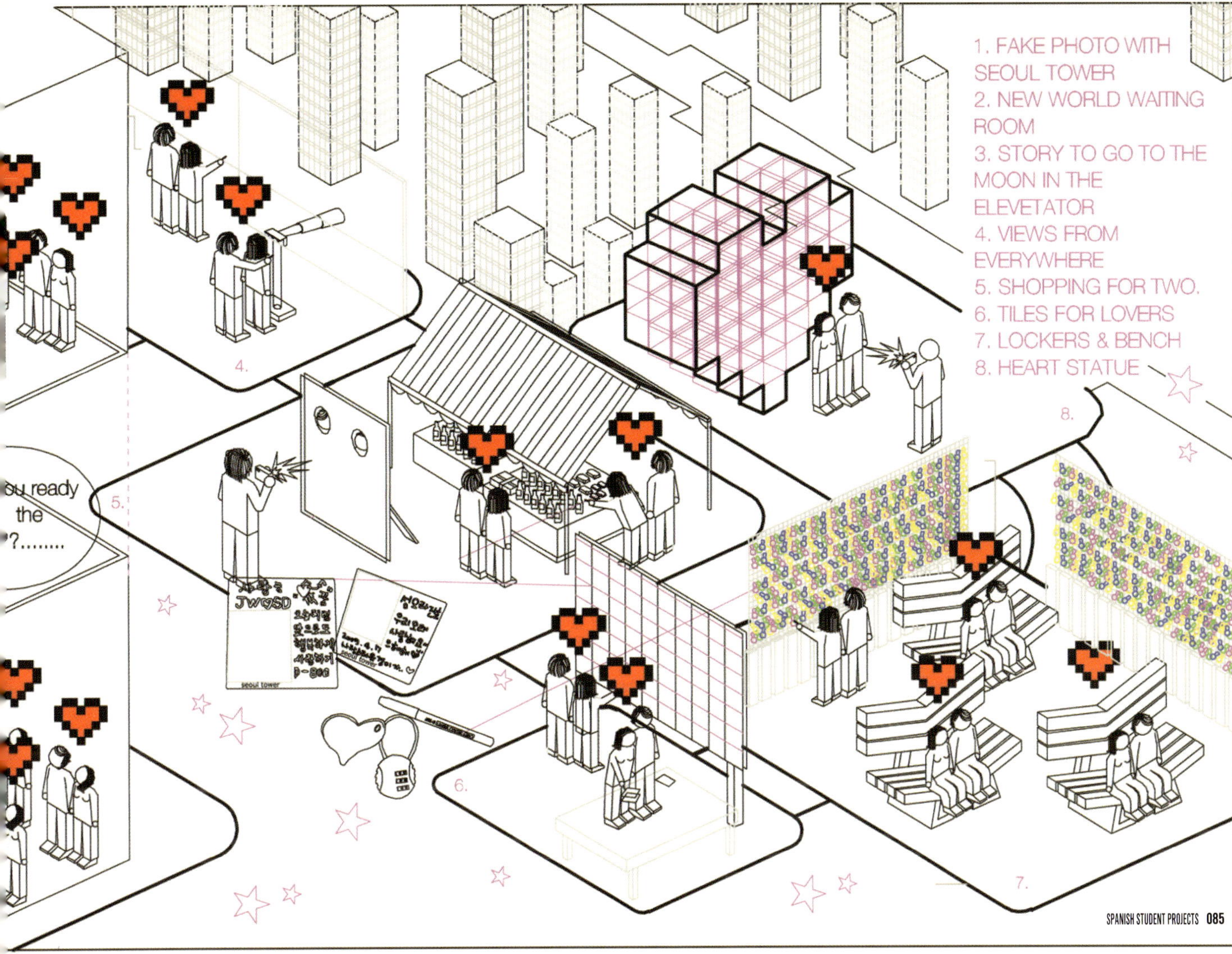
1. FAKE PHOTO WITH SEOUL TOWER
2. NEW WORLD WAITING ROOM
3. STORY TO GO TO THE MOON IN THE ELEVETATOR
4. VIEWS FROM EVERYWHERE
5. SHOPPING FOR TWO.
6. TILES FOR LOVERS
7. LOCKERS & BENCH
8. HEART STATUE
4.
5.
6.
7.
8.
ou ready
the
?........
JW♡SD
seoul tower

Hanok. Vernacular Traditional House

From a house to a Restaurant

Hanok

Casa Típica Coreana

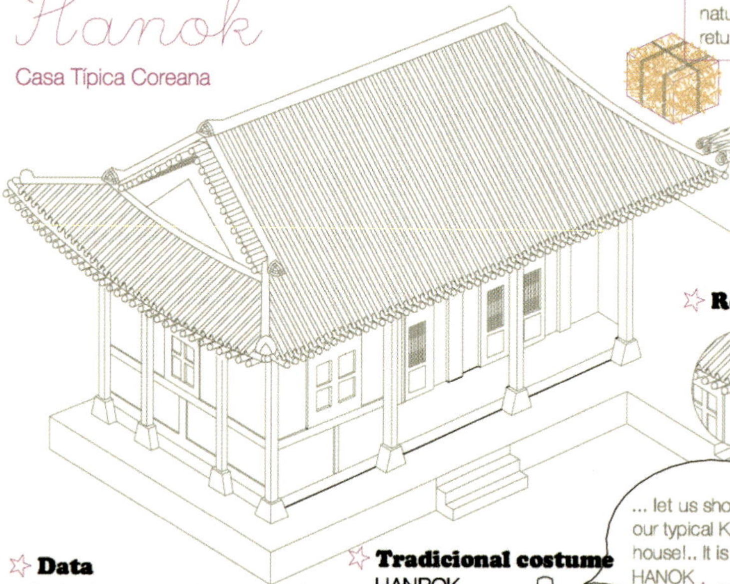

☆ **Materials**

it's made of natural components for natutal benefits. After demolition they return to Nature

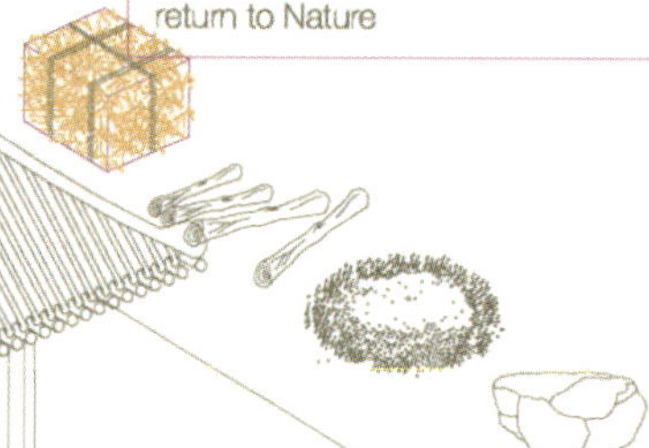

☆ **Roof. Economical Status.**

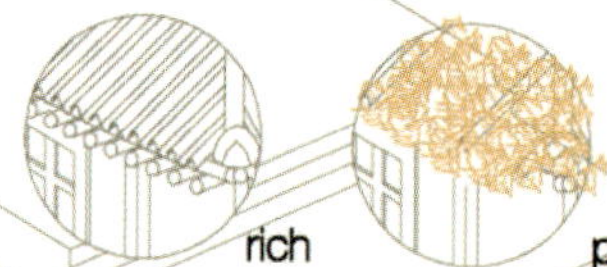

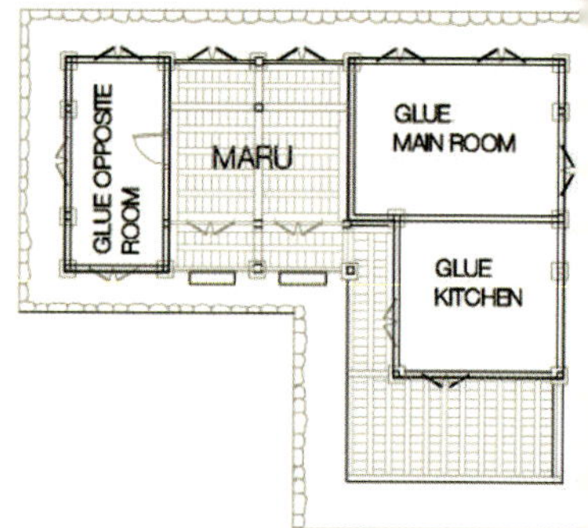

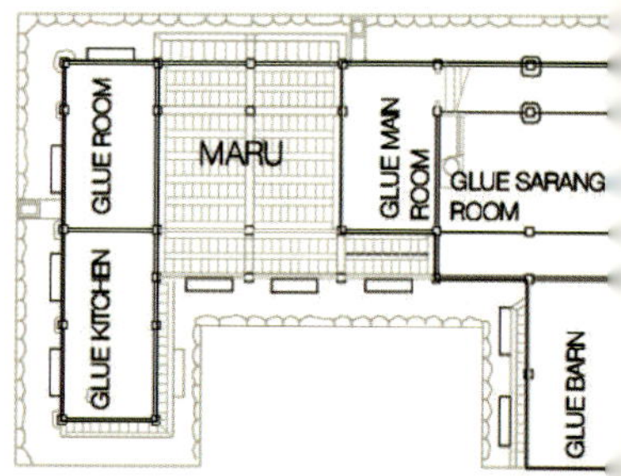

☆ **Data**

Latitude: +36.79 (36°47'24"N)
Longitude: +126.98 (126°58'48"E)
Time zone: UTC+9 hours
Country: South Korea
Continent: Asia
Sub-region: Eastern Asia
Distance: ~10000 km (from Spain)
National Flower: Mugunghwa
Korea's national anthem : Aegukga

☆ **Tradicional costume**

HANBOK

... let us show you our typical Korean house!.. It is called HANOK

☆ **Formal Hanok Shapes**

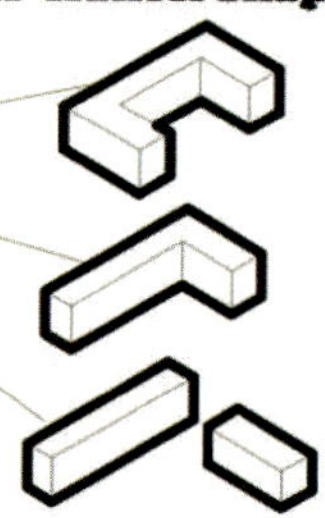

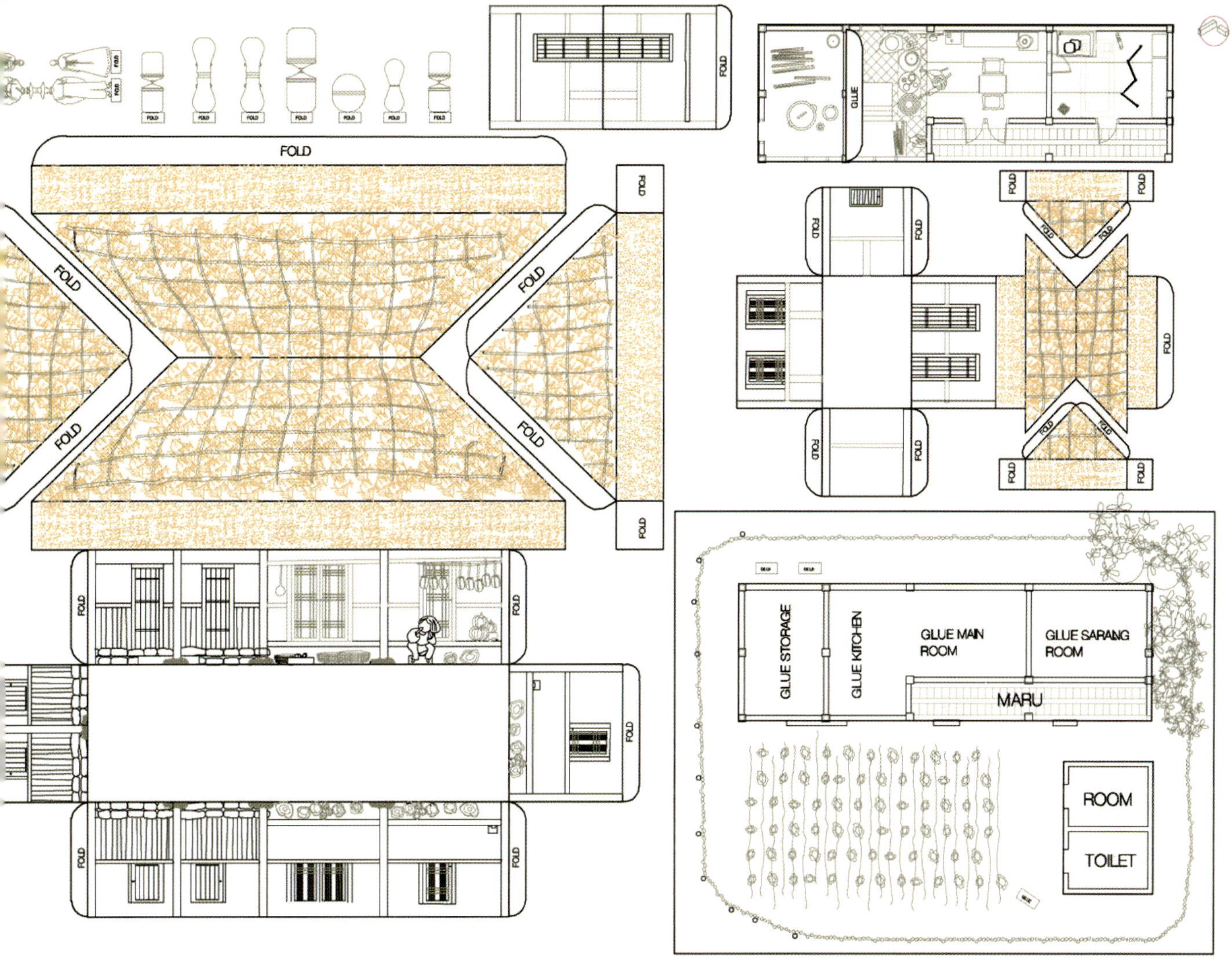

FOLD
GLUE
GLUE STORAGE
GLUE KITCHEN
GLUE MAIN ROOM
GLUE SARANG ROOM
MARU
ROOM
TOILET

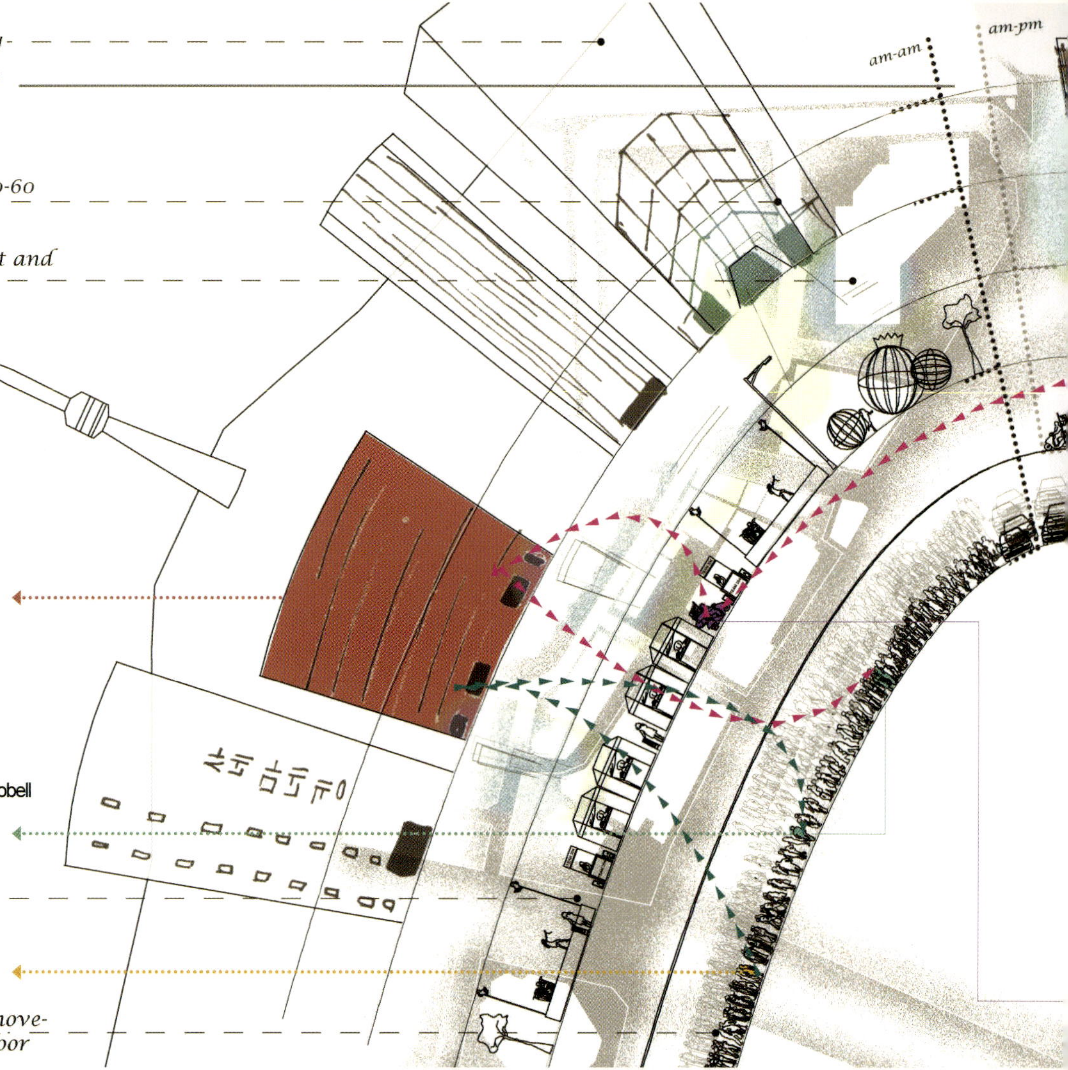

Belen Gomez Llobell

Dongdaemun Market

This area of Seoul has grown in a completely different way than other commercial areas. From the very beginning , it was a wholesales market that was opened in the night from 1:00am till 1:00 pm, and from its enormuos activity and from the growing of the city, it appears a new reatil market in the area. But as its origins are so different the whole urban structure works in a completely different way compared with other commercial areas in other cities or even in Seoul.

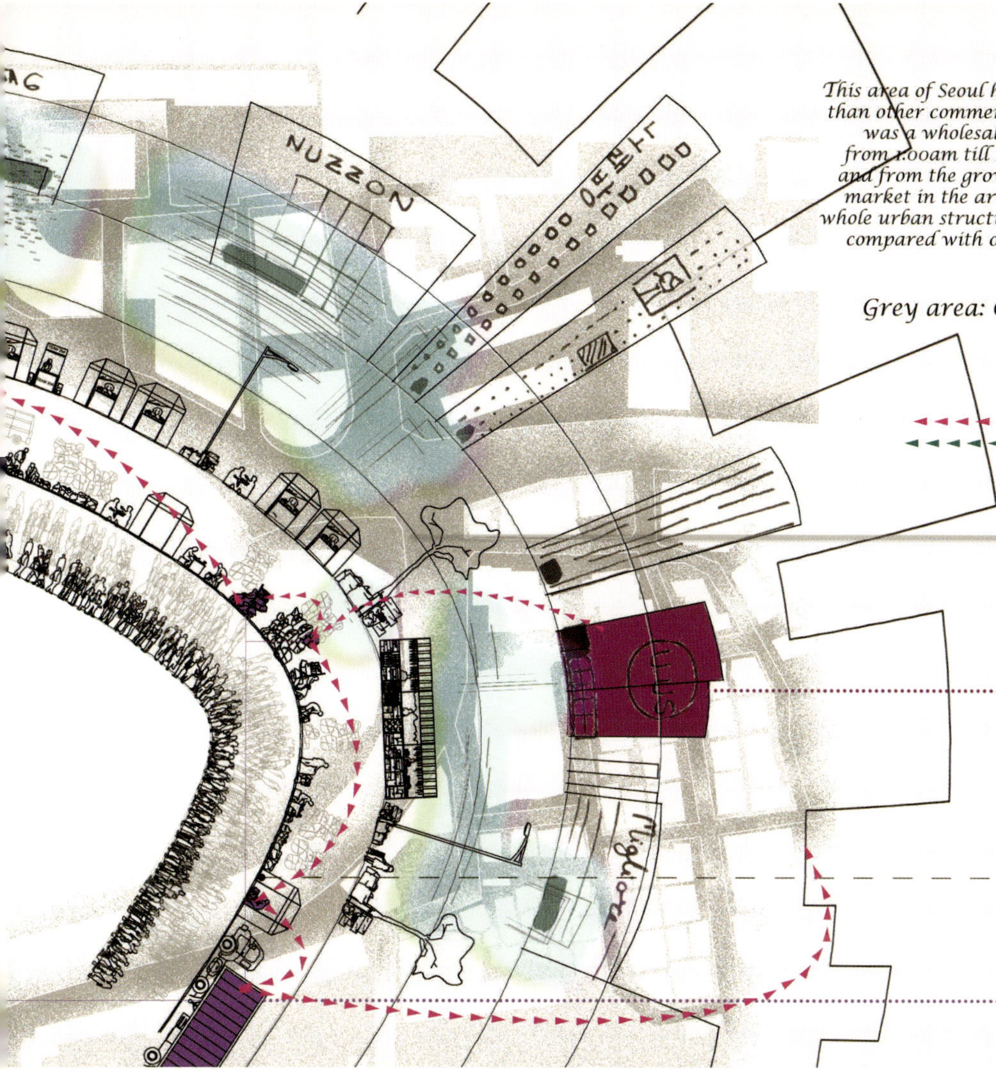

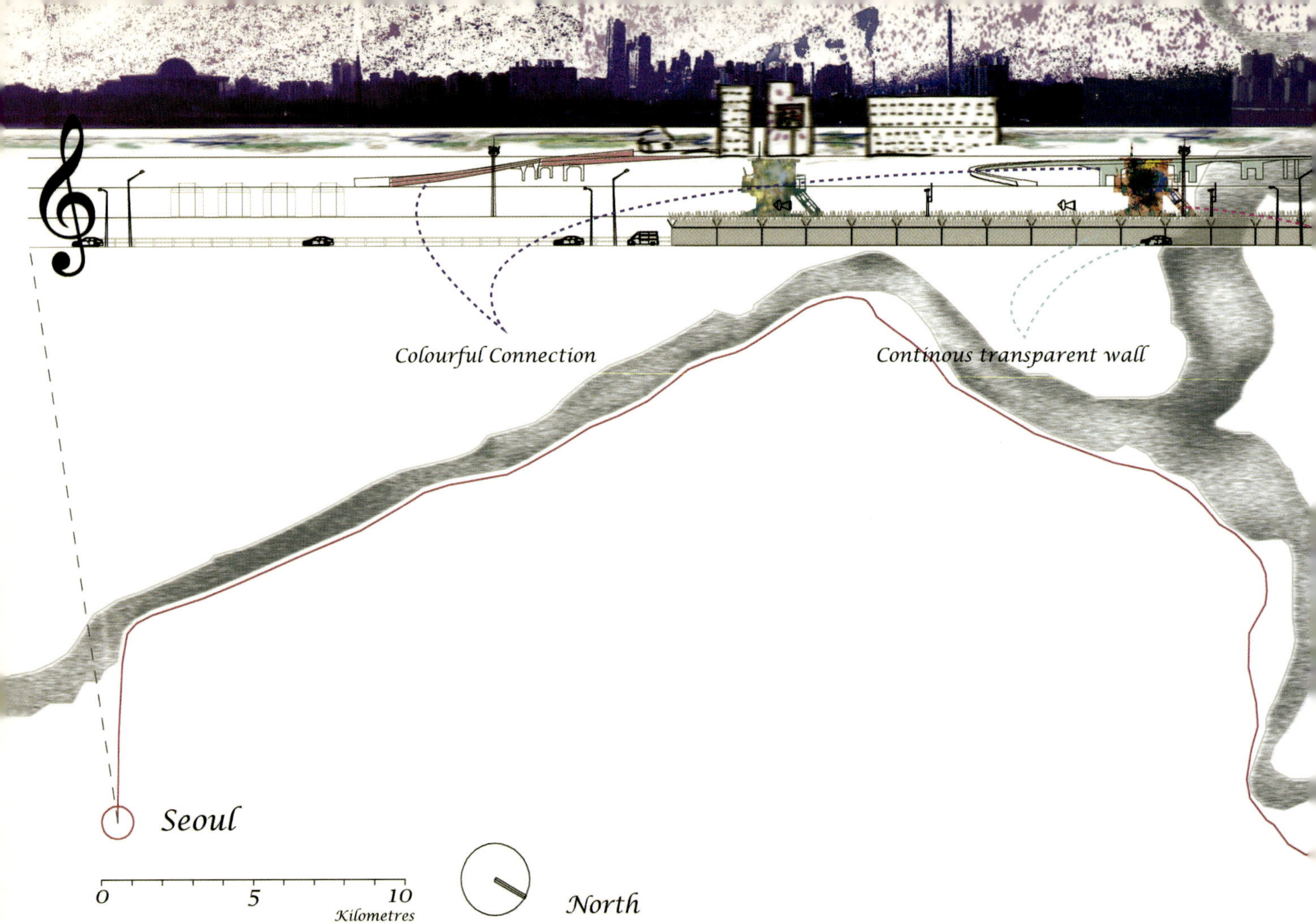

Colourful Connection
Continous transparent wall
Seoul
0
5
10
Kilometres
North

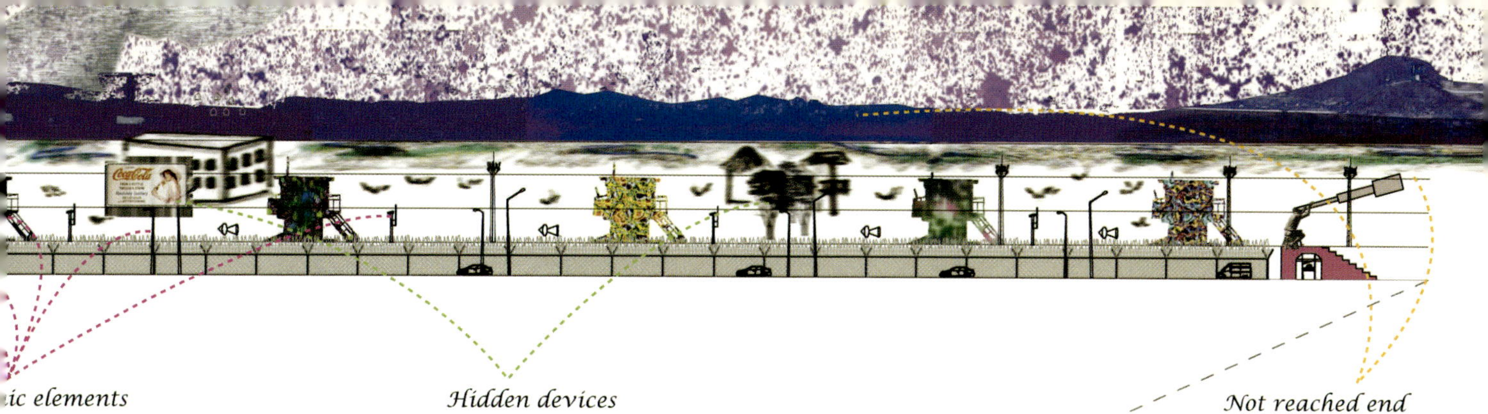

Landscape Elevation

Scenic road towards North Korea:

The scenario appears meanwhile we drive toward Imgingak peace park. The military area peeps out behind the transparent fence.

Some elements are repited along the road providing a rytmic pattern to the elevation. Military stands, speakers, antennas, acceses and lamp post.

As we get closer the scene becomes more interesting, posts and trees try to hide different military buildings, but are never completely hidden.

With another mechanism but creating the same attaction buildings of North Korea appear in the distance.

By different mechanisms, it is created a touristic military fence.

Imjingak peace park_Tourism Organization

63 kilometres

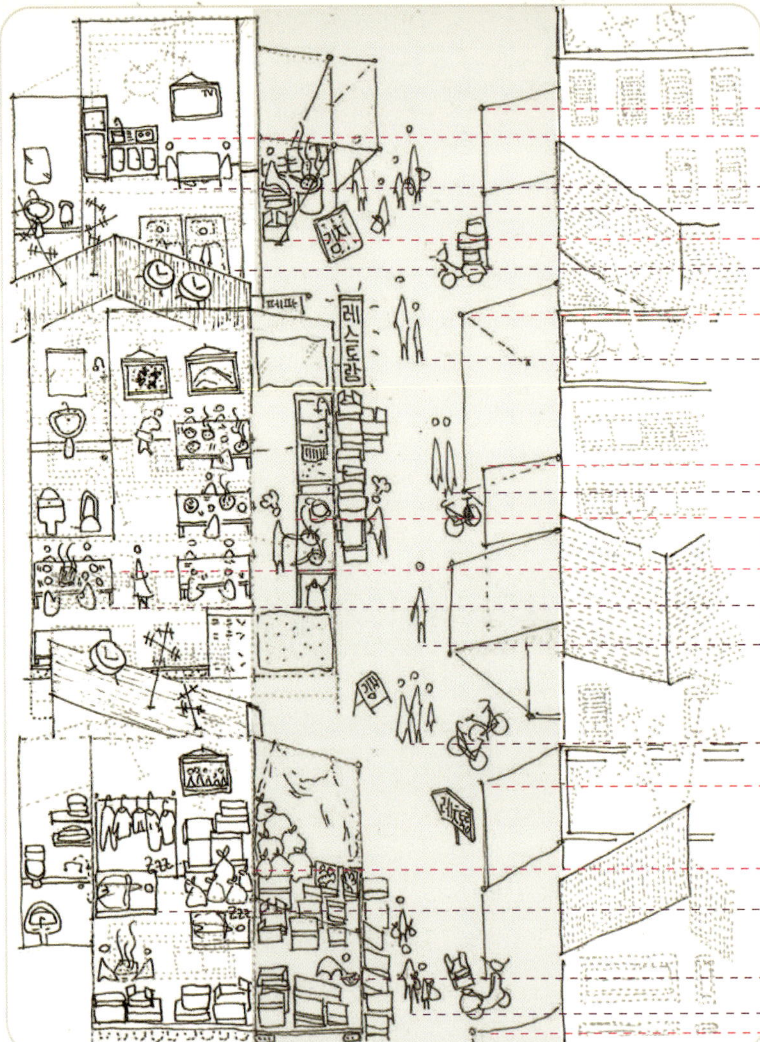

레스토랑

i would like two of these and one of those ones
bla, bla...
omg! there's too much noise outside... they won't let the kids sleep...
what would you like to eat?
we want soju for all, and kimchi and bulgugi as well
bla, bla...
it is cold outside, tonight!
Z z z z...
Mrs Park, how is your family?
i can see that Juhye & Hyemee are sleeping
bla, bla...

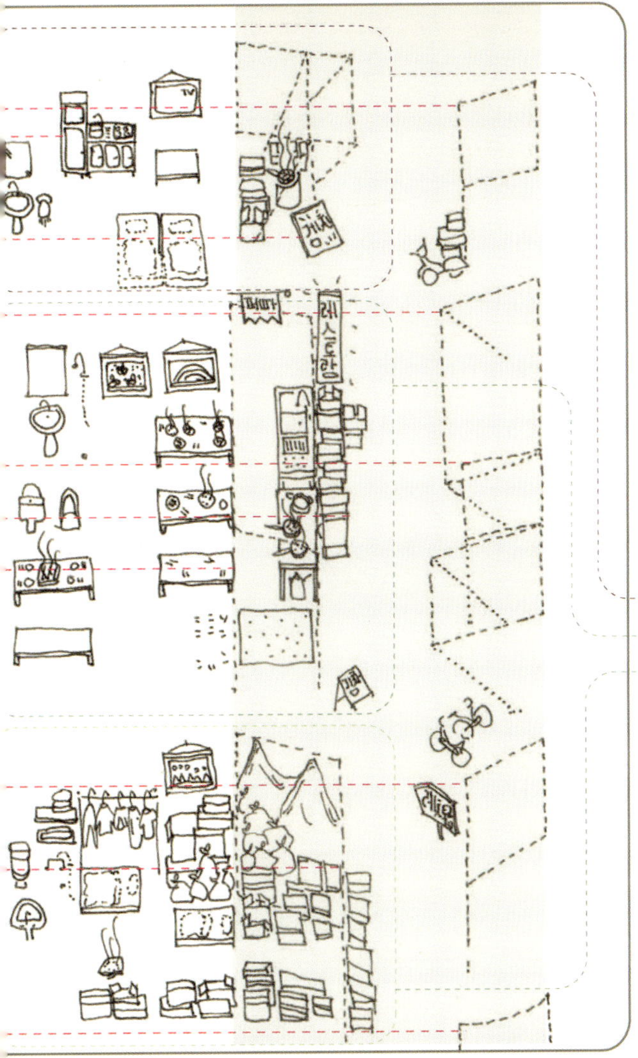

Mary Pepa Ramos Mira

In the area of Saechon, specifically in Jahamunro street, we detect a very appealing situation that is quite common around Seoul.
The interesting in/out relationship, in an 'urban market' situation; established by the strategy of public space [street] appropiation. The buildings extend beyond their private border to the public domain.

Socks selling light truck, dwelling with food selling porch, pipes store closed with all the stock in the street, restaurant with an outdoor kitchen, dwelling with independent woman occupying [without permission] a fragment of the facade with her little convertable stand, ... a series of different in/out relationships that take place but that can be embraced by 3 categories.

*case 1_parasiting preexisting buildings
*case 2_related to catering trading
*case 3_related to the domestic domain

On one hand, it exists a direct and visual in/out relationship in social terms. Users, workers and owners are connected in the same level.

On the other hand, the uses of known spaces are redefined, and the thick line between the buildings and the streets gets dissolved, to the point that you cannot longer point out, precisely, the limits of the private and public domain.

1. Noryangjin fish wholesale Market

KILL FISH

2. Gwangjang Market

SEAT 'N EAT

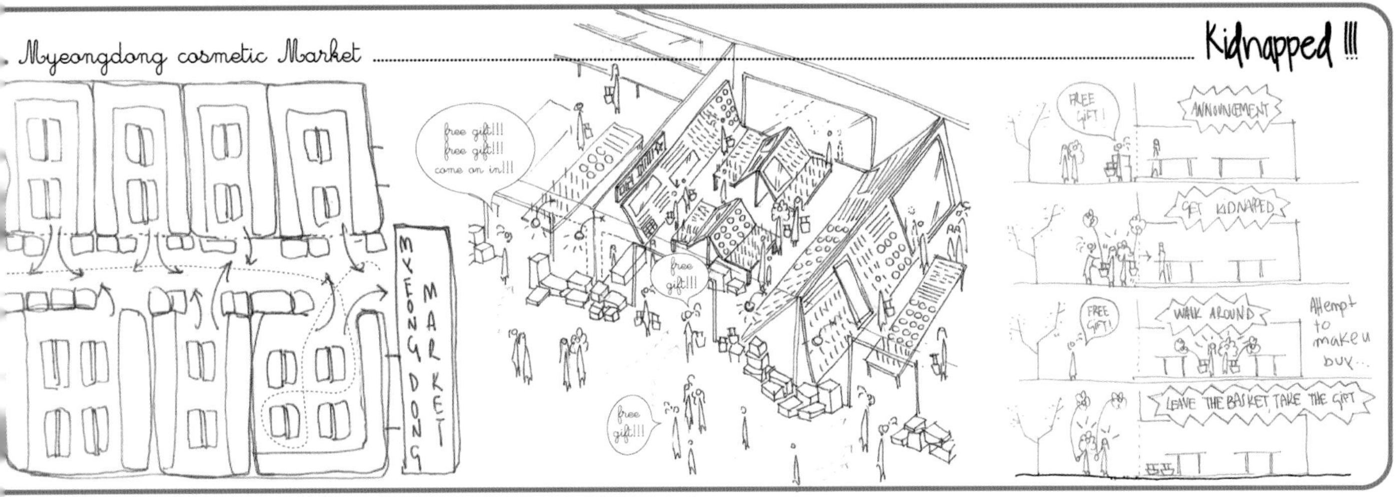

he different specialized markets around Seoul, have their own specific conditions according to the stock they are se-ng, but there are also strong common features.

he main difference is the direct and visual relationship between the user and the product. In some cases from the be-ining till the end there is a constant connection while in other cases, there is an in between step before the delivery of

ach stand, as well as in the urban markets, has a need of expanding itself beyond its limits. This provoques a con-t with the stands nearby, in a way that there is no visible difference on which product belongs to which stand.
he imagery and signage of each shop, looses strength against the amount of products, boxes and stuff to sell. These e the ones that set up the space. The architecture plays a very sutil role, that allows as a kind infraestructure, that all se uses take place.

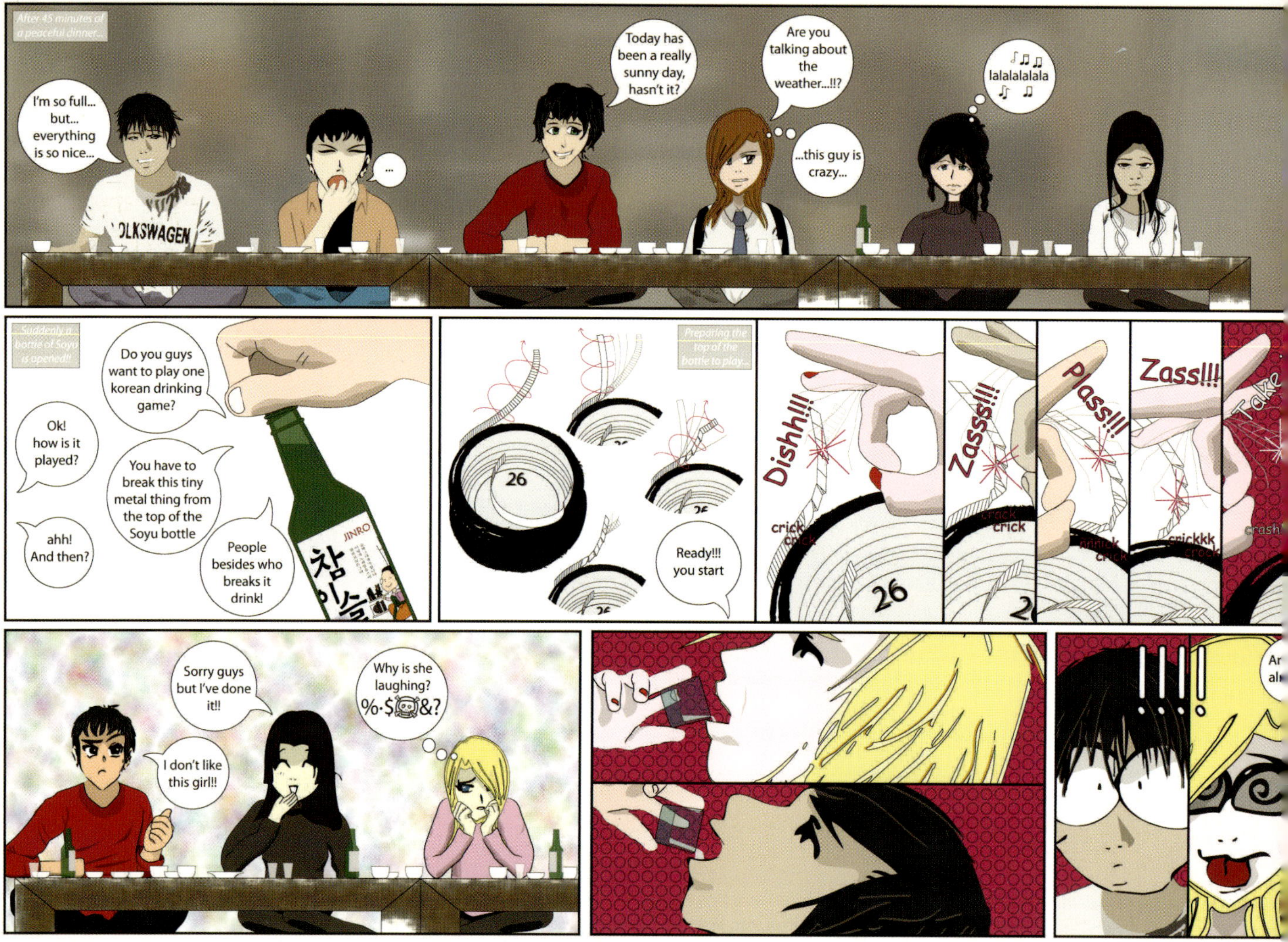
After 45 minutes of a peaceful dinner...
I'm so full... but... everything is so nice...
OLKSWAGEN
...
Today has been a really sunny day, hasn't it?
Are you talking about the weather...!!?
...this guy is crazy...
lalalalalala
Suddenly a bottle of Soyu is opened!!
Do you guys want to play one korean drinking game?
Ok! how is it played?
You have to break this tiny metal thing from the top of the Soyu bottle
ahh! And then?
People besides who breaks it drink!
JINRO
참이슬
Preparing the top of the bottle to play...
26
Ready!!! you start
Dishh!!!
crick crick
Zasss!!!
crack crick
Plass!!!
nnnick crick
Zass!!!
crickkk crack
Take
crash
Sorry guys but I've done it!!
I don't like this girl!!
Why is she laughing? %·$&?
!!!!

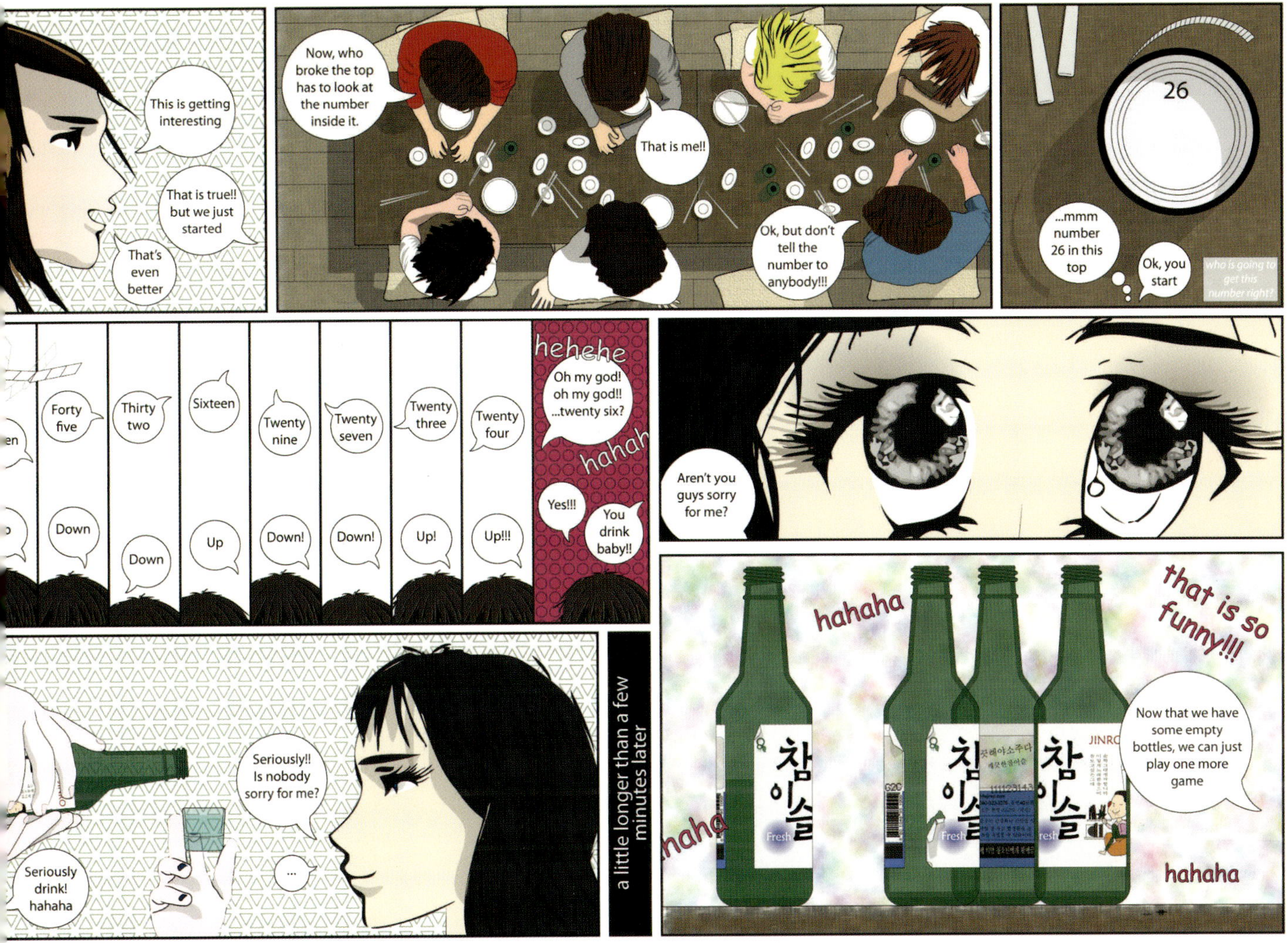

This is getting interesting
That is true!! but we just started
That's even better
Now, who broke the top has to look at the number inside it.
That is me!!
Ok, but don't tell the number to anybody!!!
26
...mmm number 26 in this top
Ok, you start
who is going to get this number right?
Forty five
Down
Thirty two
Down
Sixteen
Up
Twenty nine
Down!
Twenty seven
Down!
Twenty three
Up!
Twenty four
Up!!!
hehehe
Oh my god! oh my god!! ...twenty six?
hahah
Yes!!!
You drink baby!!
Aren't you guys sorry for me?
Seriously drink! hahaha
Seriously!! Is nobody sorry for me?
...
a little longer than a few minutes later
hahaha
that is so funny!!!
haha
참이슬
JINRO
Fresh
Now that we have some empty bottles, we can just play one more game
hahaha

meanwhile a girl starts to flirt with...
Well... I'm going to spin the bottle and who is pointed will be asked by one of us
Any kind of question?
Any
Then?
Who doesn't answers the question, drink!!
This game sounds as interesting as you...
hahaha
hahaha
haha
Fiufff...
Yes!!
Nooooo!!!
hahaha
Oh my god!
Don't be bad with the question...
Have you ever drunk Soyu with your parents?
It's so soft...
Of course!!
Next question more racy!!! please...
Who is gonna be the next!!?
Upsss..
Have you ever taken a picture to your boyfriend naked?
...
drink!!
drink!!
drink!!
drink!!
drink!!
drink!!
drink!!
drink!!
hahaha

SOJU: MECHANISM OF SOCIALIZATION

Soju bottle acquired in this circumstance the architectural skill of mechanism of game. This mechanism generates situations of socialization in a context designed as highly domestic space. The top, with its system of openness and random number on the back, and the bottle itself enable a series of horizontal relationships between the companions at table.

Interpersonal hierarchy that exists in Korean culture (based on age or position held in a given environment) is reflected in every daily situation. However, the ritual that is created through the bottle of Soju generates a certain horizontal social role among participants. It establish a much more fluent and sincere environment. Therefore, it generates less static and inhibited situations caused by the range of the people.

As soon as the games are going on, the capacity of extroversion of each individual increases. Moreover the degree of affinity among the participants raises too. These are caused by the consumption of Soju inside the game process. These kinds of processes always occur in closed and private environments where the public context disappears. To create a similarity with Spain, we can say that it would be impossible to see situations close to our botellon in Korea. In Spain the situation occurs anywhere in a city, anytime in a city, whether public or private.

Alejandro Chichizola Olivieri

KOREAN JJIMJILBANG - MULTIFUNCIONAL BUILDING

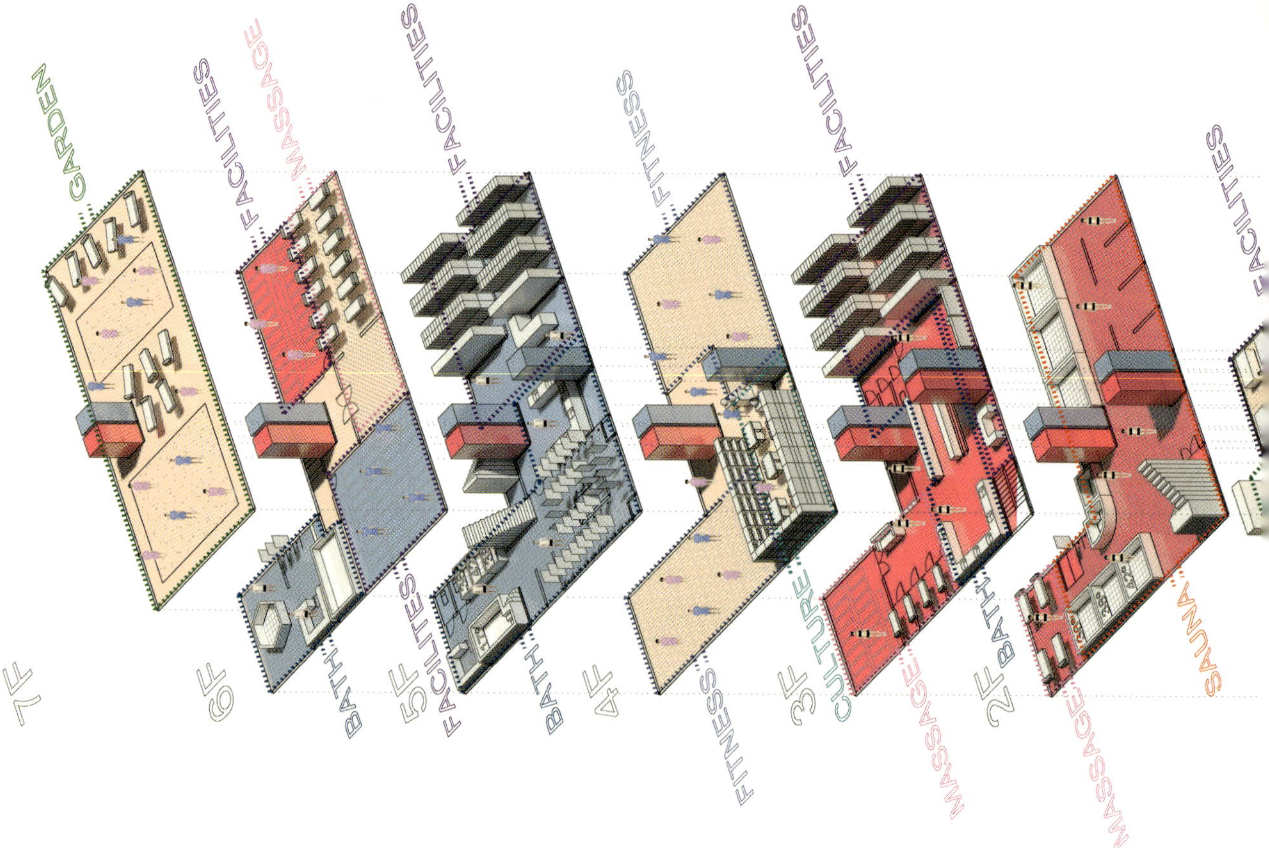

Eduardo Landete

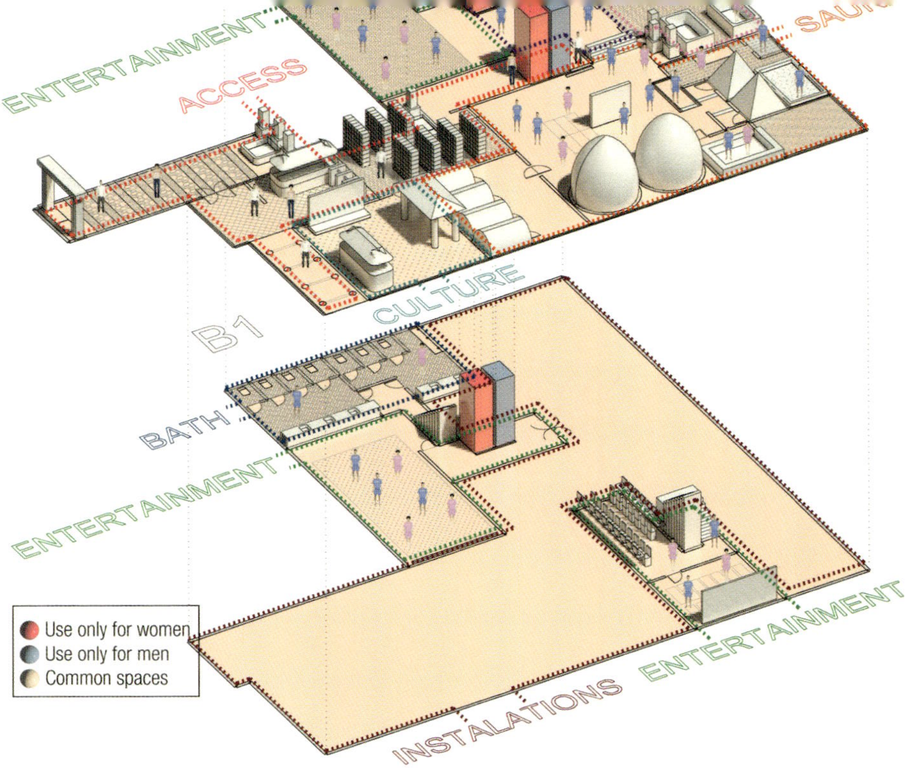

DRESS PROTOCOL

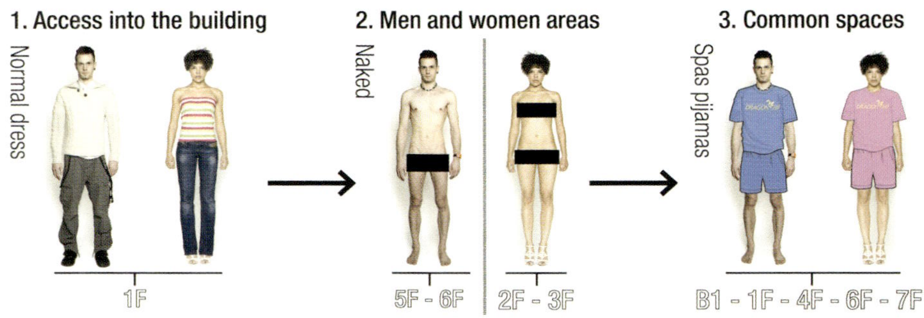

Complexity of the Seoul Streets

서울시 강동구 성내St.

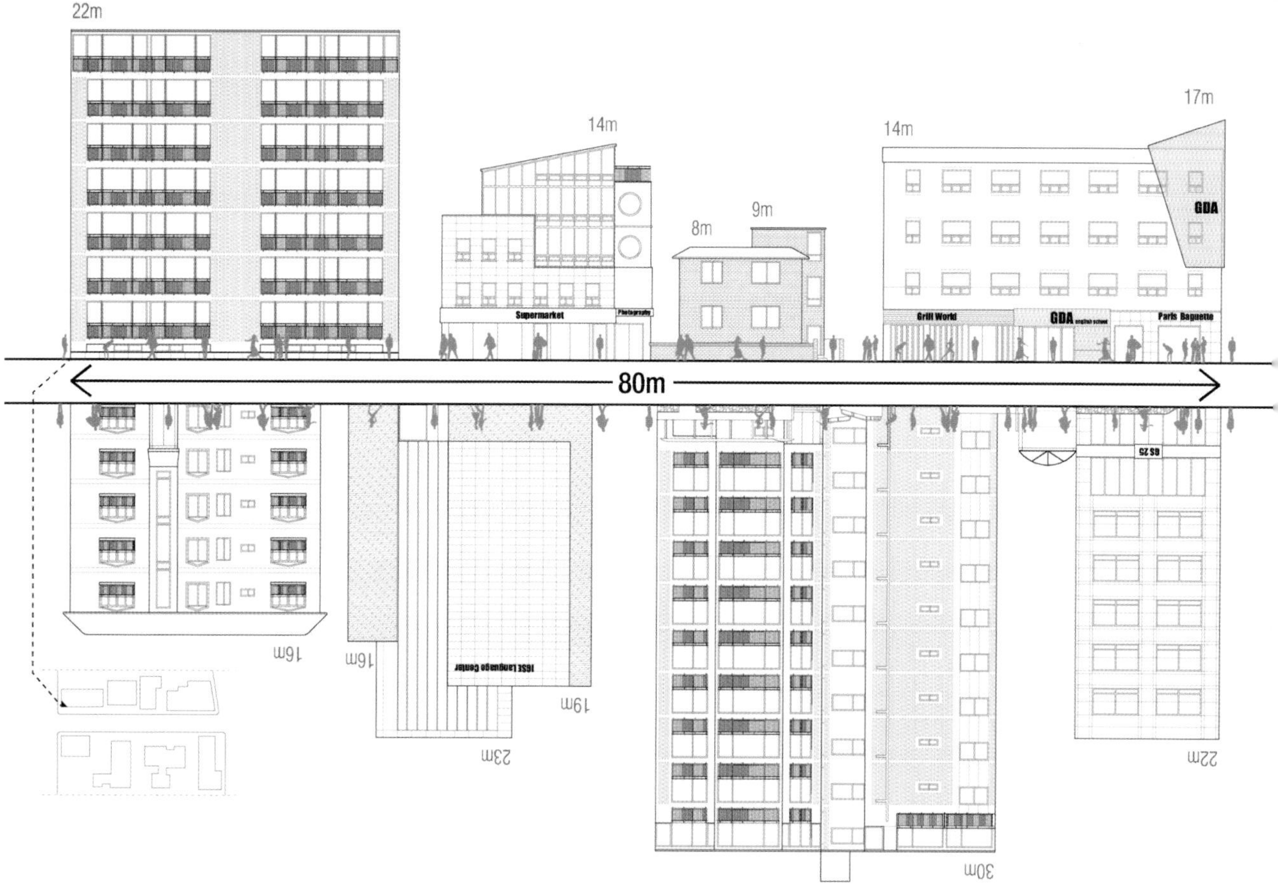

Seoul is a city of contrasts. That's something that we realized the firsts days. Just in one street we can find many different programs. There are not rules in the way of building architecture. Because of that we have chosen a little street of 80m long and we have analyzed the eight different building that set it up. So, we can detect from residential multi-family housing use to a university, passing through a kindergarden or a residential family house. The complexity of this program in just one street is really deep. There are many different users along the day. The street is used during the whole day for many people in different ways.

There is people in the morning leaving the kids in the kinder garden. People going to the university during the morning. Eating in the restaurant in the midday. Going to the supermarket during the afternoon and buying stuff in the 24h market during the night. So, in just 80 meters we can find eight different buildings with seven different typologies. And its repercussion within the use of the street. It's a situation that is repeated along Seoul and we consider that is the typical way of understanding the city. This street is just an example, but Seoul is a 24h city where everything can happen anytime.

María Foulquié

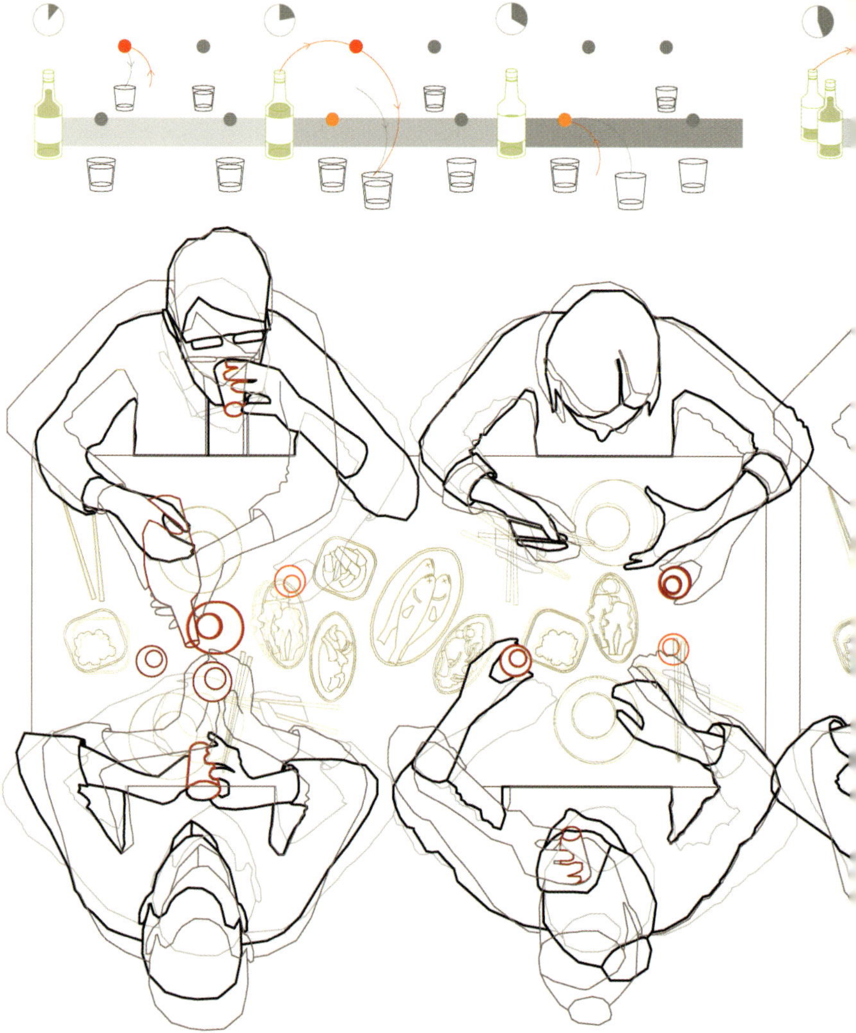

Drinking Soju

Drinking Soju is an event, a ritual, a choreography.
It is a ritual as a set of actions of a symbolic value, based in a belief system. When you sit down to share the Soju you do it to build a relationship of trust and friendship around the idea of sharing food, alcohol, and confidences. Drinking from the same cup is the symbol and also the foundational fact of this relationship of trust, and you do it in the shared faith on the power of alcohol as a social precipitator.
This ritual has a specific choreography, a sequence of movements that express attention to social hierarchy and a sense of respect legacy of Confucian thought. The senior opens the dance by emptying his glass and offering it to anyone who wants to entertain. He will hold it with both hands as a sign of respect while being served, and after drinking the Soyu will return it to its original owner. This sequence will be repeated with the other guests, shared glasses will multiply, and as Soju bottles are emptied, the dance will lose stiffness and improvised movements will appear. Hierarchy fades while the evening progresses.
This progressive chaos, the anecdotes that are shared or originate in the table are those that make the ritual into an event, an affair that causes a interruption in the normal course of events, and is worthy of being remembered.

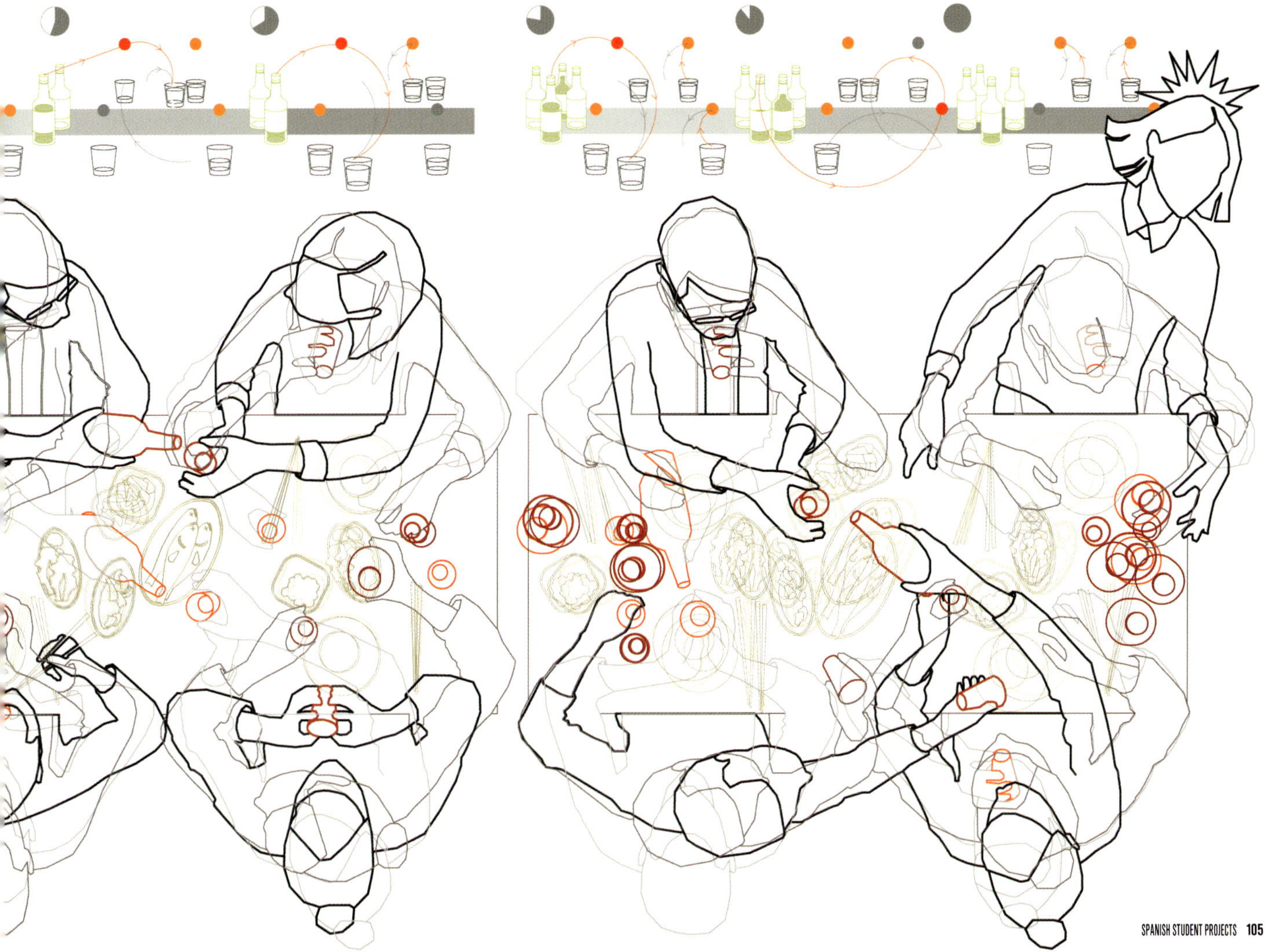

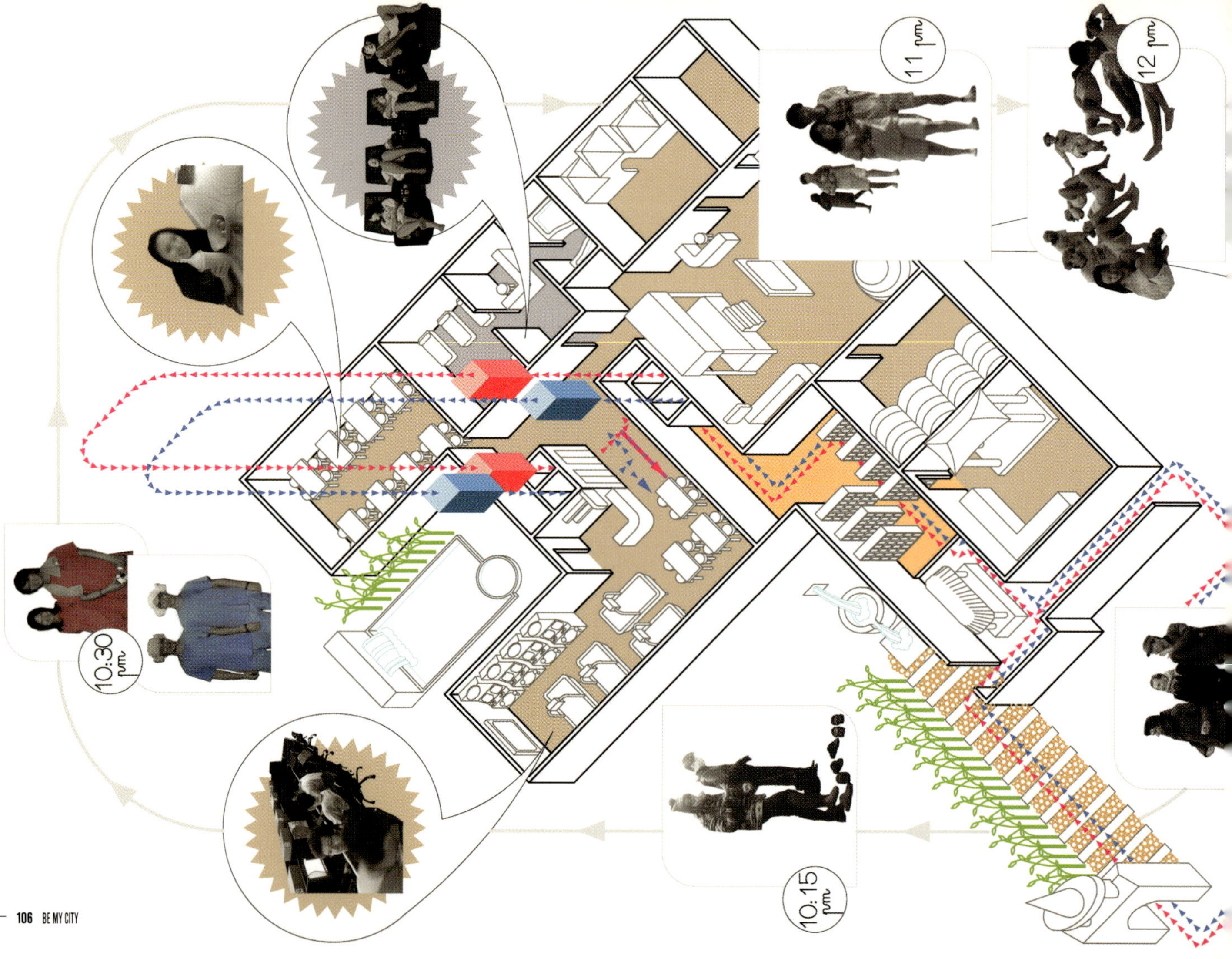
10:30 pm
10:15 pm
11 pm
12 pm

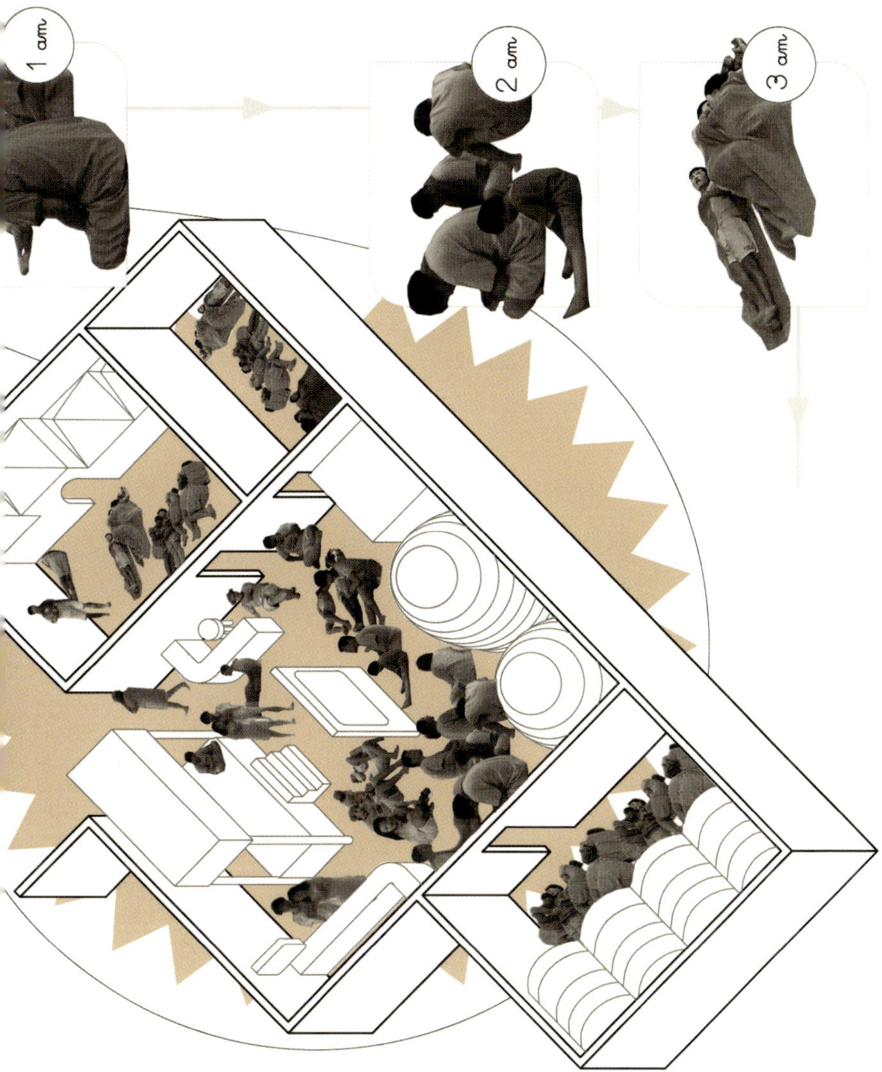

Public domesticity in Jimjilbang

One of the more striking aspects of Jimjilbang is the variety of situations of domestic character that occur in the collective space. Here, the variety of programs in the building overlap in an even greater density: restaurants, spa, sauna, game machines and computers, and especially a large central space, that by using simple mechanisms becomes your own living room. These mechanisms are simple but effective: the pillow, the blanket, the pyjamas.

After using the public baths segregated by gender, users gather in the communal space, and reproduce scenes typical of the privacy of the home. Eating, playing, resting, chatting, watching TV ... all of these situations occur in the community, surrounded by crowds of strangers whose presence is blurred due to the uniformity that provides pajamas.

As the night goes on people goes to sleep anywhere, side by side, crowded into the rooms dedicated to it or, if no other choice, in every corner sheltered from the lights, the noise of the television and people still awake, or the activity of the cleaners doing their job.

In this way the Jimjilbang is more than an expanded spa with other leisure programs, is also a unique way to practice public domesticity.

Cristina Moreno

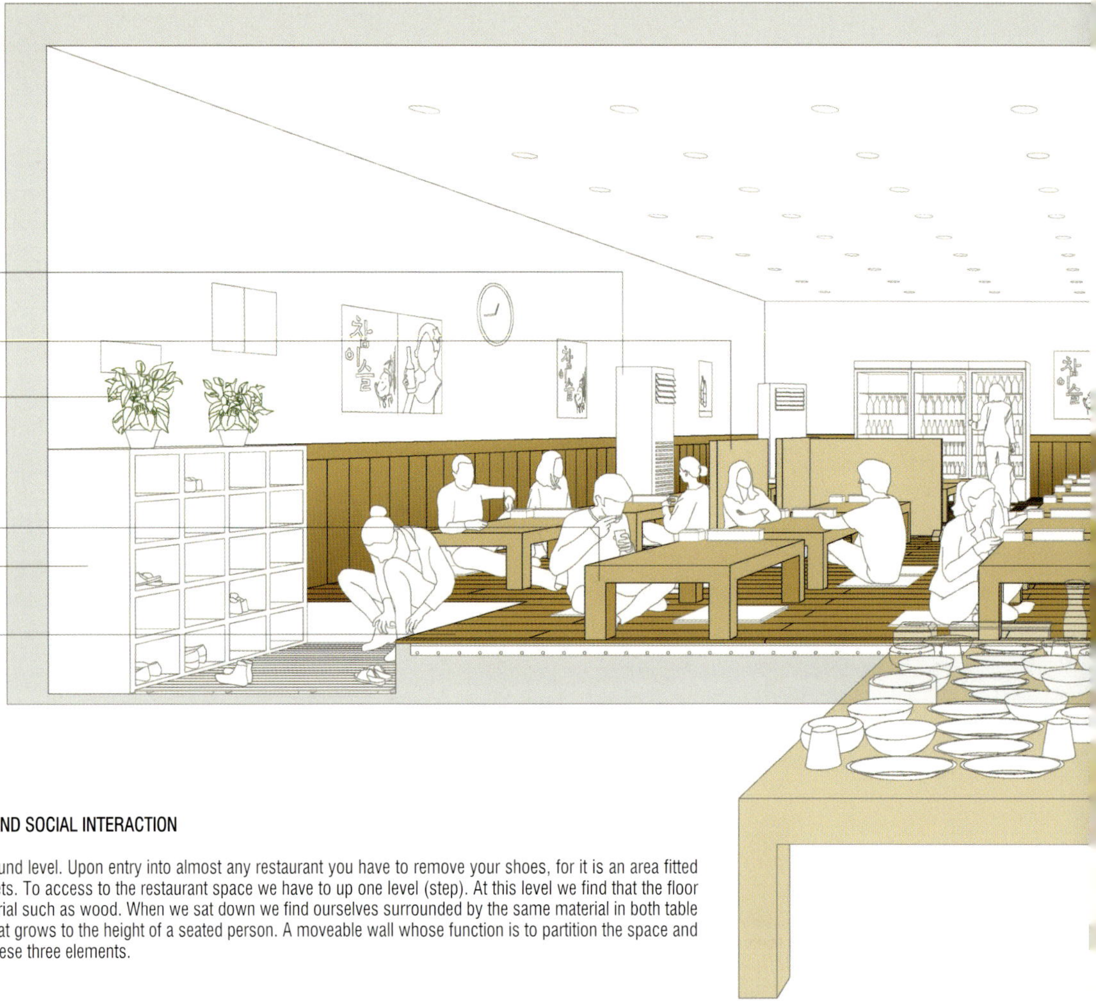

RESTAURANT AS A PLACE OF COMFORT AND SOCIAL INTERACTION

In Korean restaurants the action runs at ground level. Upon entry into almost any restaurant you have to remove your shoes, for it is an area fitted in the lowest level consisting of shoe cabinets. To access to the restaurant space we have to up one level (step). At this level we find that the floor is heated and covered by a very warm material such as wood. When we sat down we find ourselves surrounded by the same material in both table and floor and also on the base of the wall that grows to the height of a seated person. A moveable wall whose function is to partition the space and have the same height as the base joins to these three elements.

This space of comfort always complies with the same warm material (wood or derivatives). In addition to this first step in some restaurants up one level (one more step) can be found isolated compartments that can communicate with each other through a curtain moving. This interior space of the compartments is configured in the same way that the overall space of the restaurant.
The restaurant space is often a fairly large space with tables in the central space with many objects on the walls. You can find refrigerators, water machines, stoves, plants and numerous posters advertising drinks.

SUBWAY CARS AS OTHER HOME

Riding the Seoul subway was for me the discovery of a second home for the Korean people. Most people use the subway in Seoul as daily transportation for either long or short rides, moving to the public transport their daily actions. One of the characteristics I want to highlight of Seoul Subway is that users extend their hours of sleep while traveling, they choose routes with the fewest transfers possible to sleep for longer, although this ride is longer. Another situation that I found on the subway is that it is almost essential to have a Smartphone, laptop or any electronic device while traveling.

The Seoul population is in constant digital contact most of the time and while they are traveling, they use that time to talk with friends, watch a movie or tvserie, or read the newspaper online. Another important role in the subway is the heated seats as comfort level. In addition to these factors, the life underground has almost all the services needed to survive without surfacing.
The subway is full of peddlers that carry their product in small carts all over the cars. They are also striking billboards that can be found in most of the cars announcing plastic surgery.
The hierarchy of the seats on the subway in Seoul is strongly rooted in its people, as the outboard seats are for older people and people with special situations, young people never invade these seats, and older people does not interfere the rest of the car.

Gala Calatayud

Nothing is left to imagination; the external's environment construction of this noraebang is completely transparent. The facade acts as a membrane, by the light, reveals the activity that happens on the inside. The one who walks down the street is also a participant in the noraebang. Dim lights, graphic curtains, beaded lamps, stairs, and even the wallpaper's details are visible from the outside, providing a home environment. The feeling? ... It is like a dollhouse.

All such details that shape the different rooms and its relationship with the facade has wanted to emphasize in the drawing.Once inside, the home environment remains at all times. Even percieved in the diagram configuration of the interior space: a lounge at the entrance, a small garden and corridors that lead you to the different rooms of the noraebang.
Even the fact of taking off your shoes before entering, as in any domestic Korean environment, is also present here.

Inside the room, the domestic component is not leaved it up. The setting of this space makes you feel like if you were in your house living room: a large sofa, coffee table, television, curtains ...?

Though it also have another set of devices that allow the activity to occur: a small stage (as a big as a step), microphones, speakers, musical instruments, the song book, the "great" TV remote ...

Luxury 秀 노래방
토 랑
카 페 테 리 아

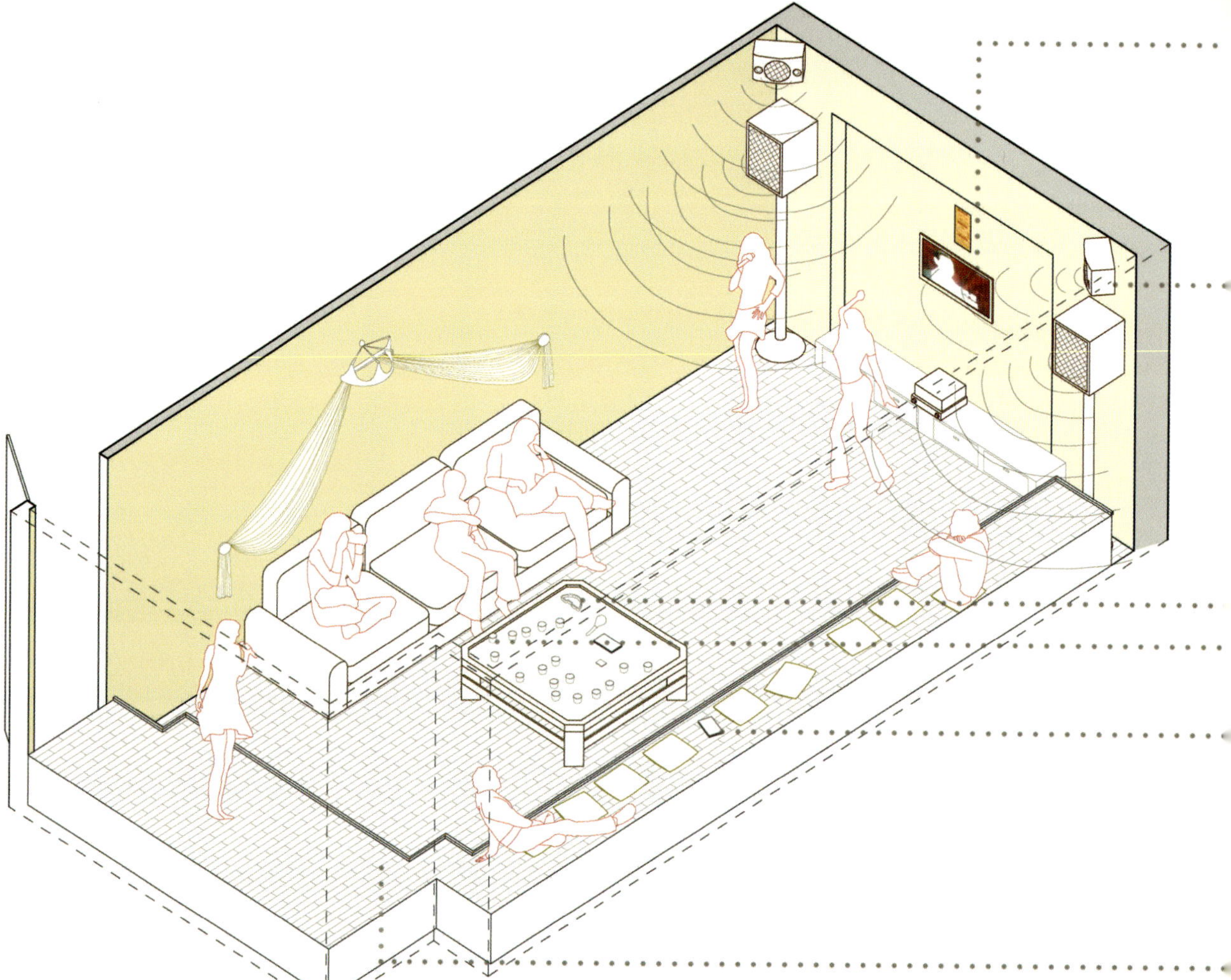

1. Television

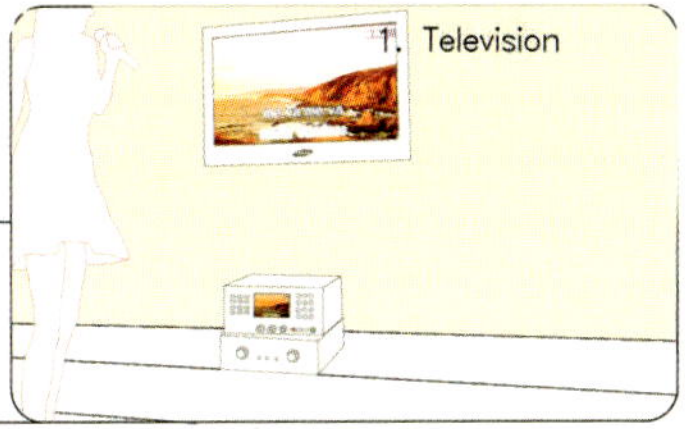

The whole room was focused on TV. Situated in the shortest wall so that everyone can see it.

2. Musical instruments

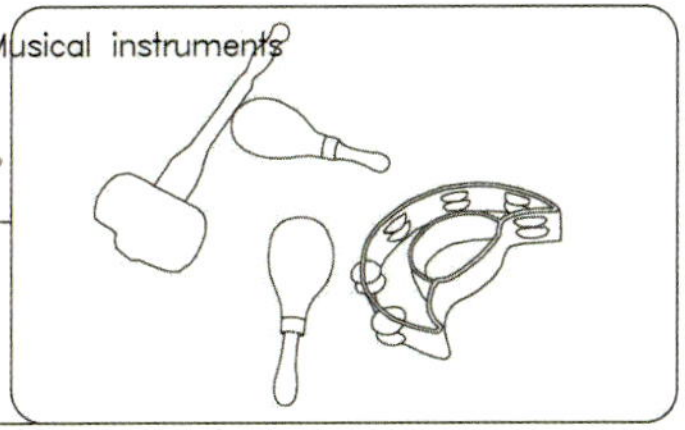

The perfect object for joining noraebang and another way of making more noise.

3. Stage

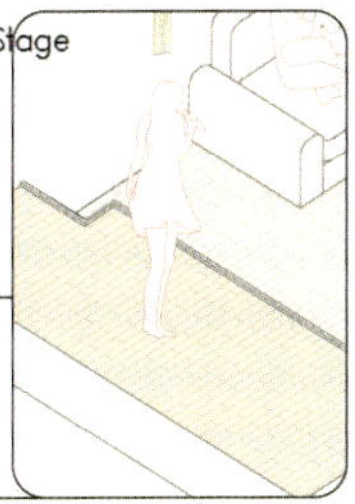

The stage in the room is constructed by a big step, just in front of the TV. So that everyone could see the performance.

4. Lights and sound

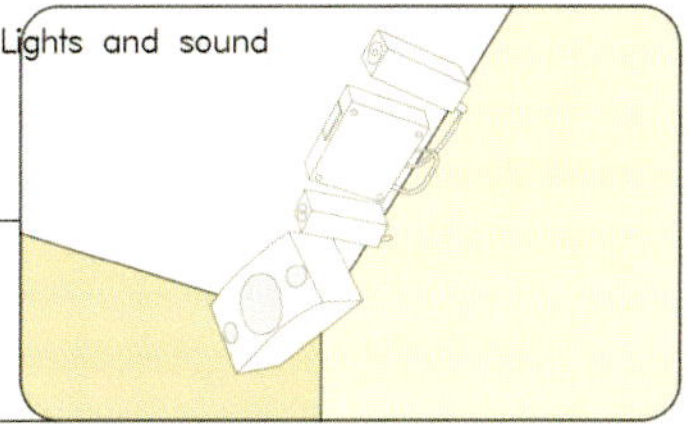

The sound devices and the low intensity of lights allow a comfortable environment in the room to start singing.

5. Song book and remote

A way of selecting your song and start to singing it!!

6. Food

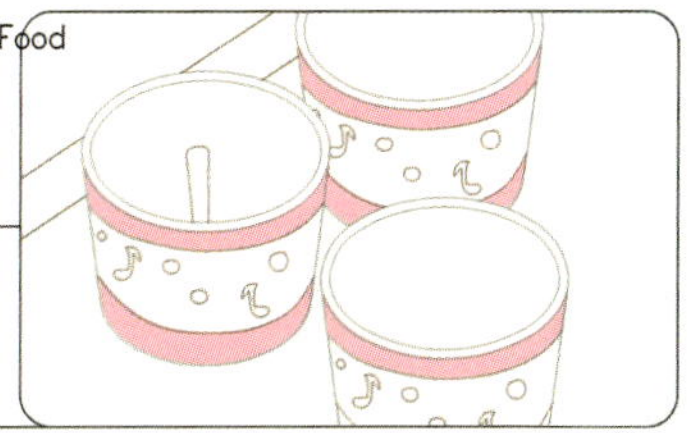

Food is also present in noraebang due to the importance in corean culture. In this case, ice-cream was free.

As domestic home's specific space, the room is where the traditional layer is mixing whit the objects en different heights. The room space is adapted and prioritized.

First of all, the drawing one shows how the first user is using the clothes hangs in a empty space; the second one shows how the bed is just been used the third one shows how the empty space is used by the second user and the first one is sleeping on the bed, an object which is the room; and the last drawing the empty space turns into usable space with just a unfolded object as the Korean bed is.

This situation shows and teaches me how all the elements of the house can facilitate using the empty-full space. How the Korean bed is folded just sleeping and it is unfolded or how the door of the room changes of position only when the empty space is used. There are lot of examples of simultaneous activities that make possible a life in a reduced space.

Adrián Beltran

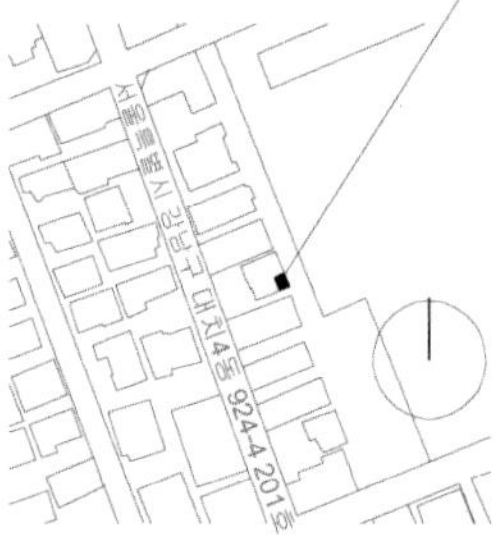

Situation 1

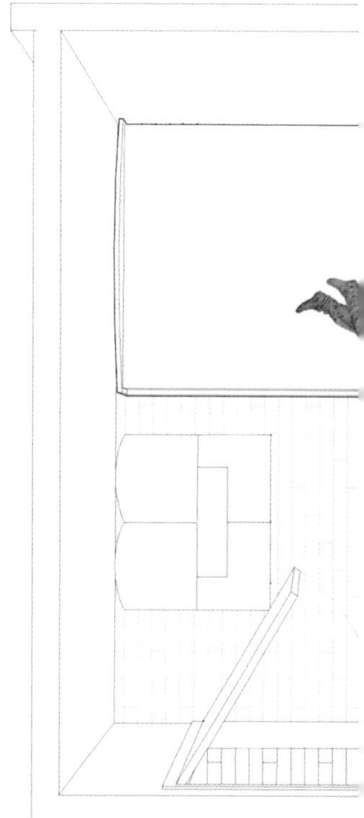

Situation 2

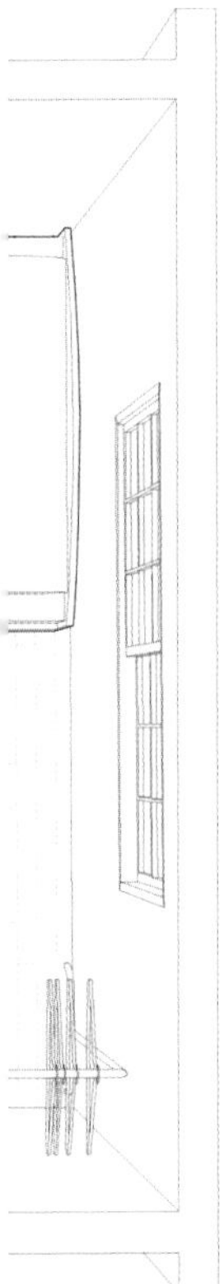

Situation 3

Situation 4

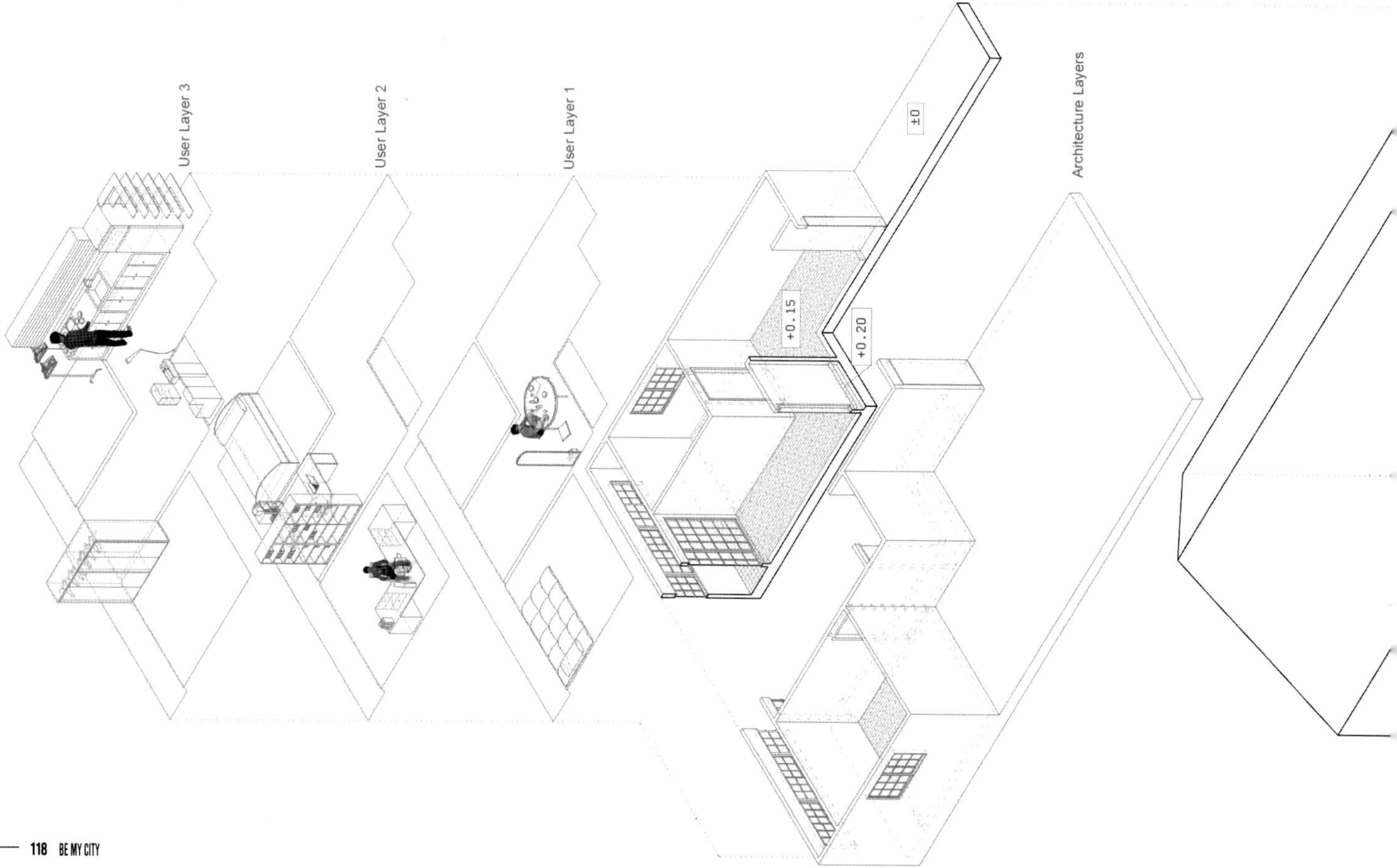
User Layer 3
User Layer 2
User Layer 1
Architecture Layers
±0
+0.15
+0.20

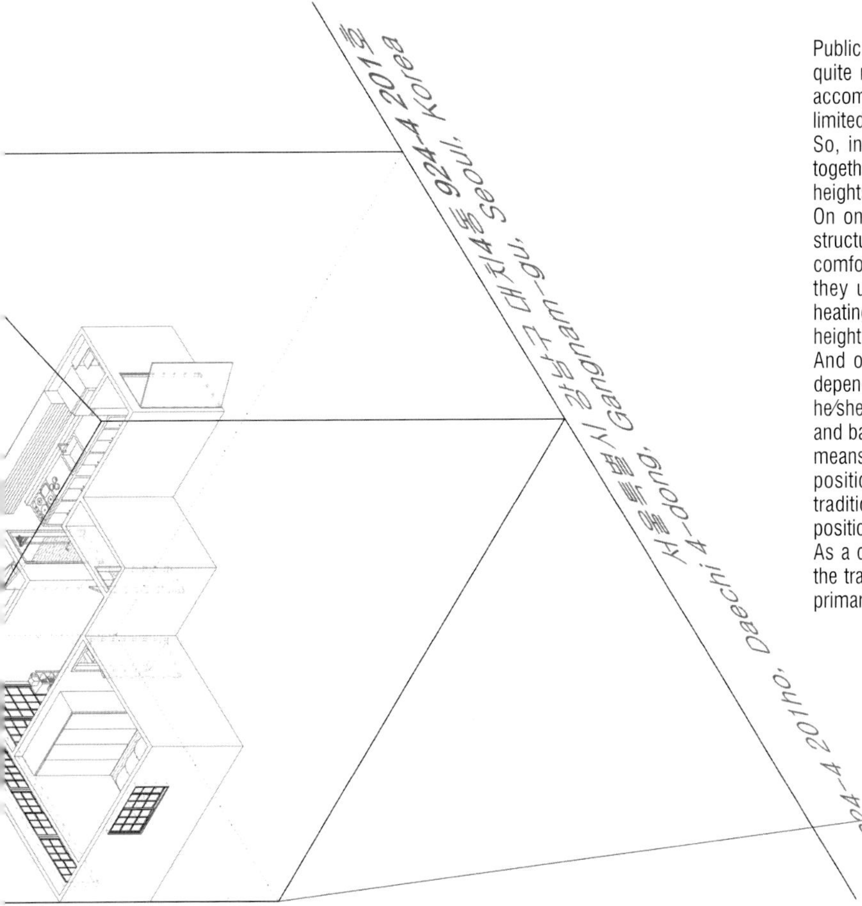

Public space in Korea is able to carry the simultaneities of uses in a space quite reduced, such it is like this that the architecture get the ability to accommodate the humans needs in a city where the space is increasingly limited.
So, in the domestic space lives the architecture and the domestic life together. As architecture as the life have 3 layers that change in different heights and the relation with the different objects.
On one hand, the first layer of the architecture is the basic one, is the structure by herself, it is the ±0 level; the next level is +0.15 that sets the comfort layer, which come from the traditional Korean architecture when they used fire and fumes, but nowadays this has changed to radiant heating; and finally the third layer is the damp layer, the bathroom, which height is +0.20.
And on the other hand the domestic life have their own 3 layers, they depend of the position of the user: the first position of the user is on foot, he/she is in this position if he is doing and unseen activity such as cooking and bathing; the second position is when the user is sitting on a chair, that means he/she is enjoying in his/her hobbies; and finally the most important position is when the user is sitting on the floor, because he is on the traditional layer of the architecture, and he is more comfortable in this position than in whatever.
As a conclusion the adding of the traditional layer of the architecture with the traditional position of the user means that the space will be where the primary activities will be done.

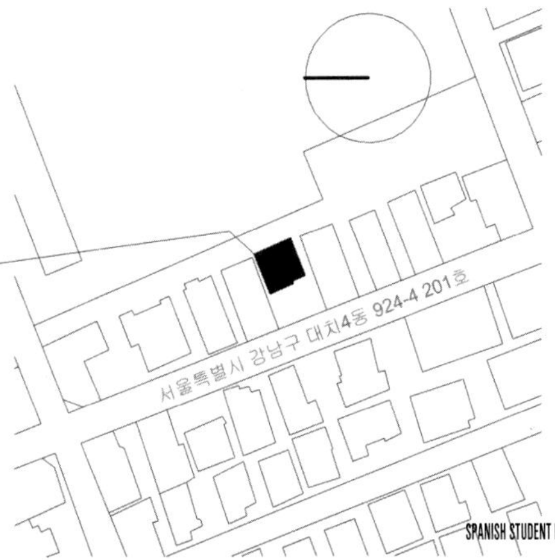

Javier Esquiva Lopez

FROM THE EXTRAORDINARY TO THE ORDINARY

The most surprising aspect was the step forward from the extraordinary to the ordinary. In just few hours I moved from a feeling of surprise, given the obvious differences between this domestic space and mine / ours (the relationship with the floor, rest, the heat, compactness…) to a feeling of familiarity. This situation was caused by the presence of an overflow of the more ordinary and personal objects. As well as my more personal and less visible space, everything was organized and placed with no aesthetic but a functional intention.

Books, cloths, bathing objects, pans, food, etc., have no determined places. They go beyond the furniture limits and occupy, according to the necessities, the most specific places like wardrobes, shelves and so transforming these into something else. The bathtub becomes a big bathing pond, the shelves for souvenirs become unexpected hangers, a telescopic structure becomes an open wardrobe… Everything is visible and shared.

Any place, any corner is valid for immortalizing any moment; the TV, above a cabinet, a shelf, a wall. The interesting thing is how everything is interlaced generating this personal and purely familiar atmosphere. It even comes to a point where customization becomes radical. A quote by Joshua, emotively revealing for Minjae, rests on his sleeping room wall.

phenomenological details

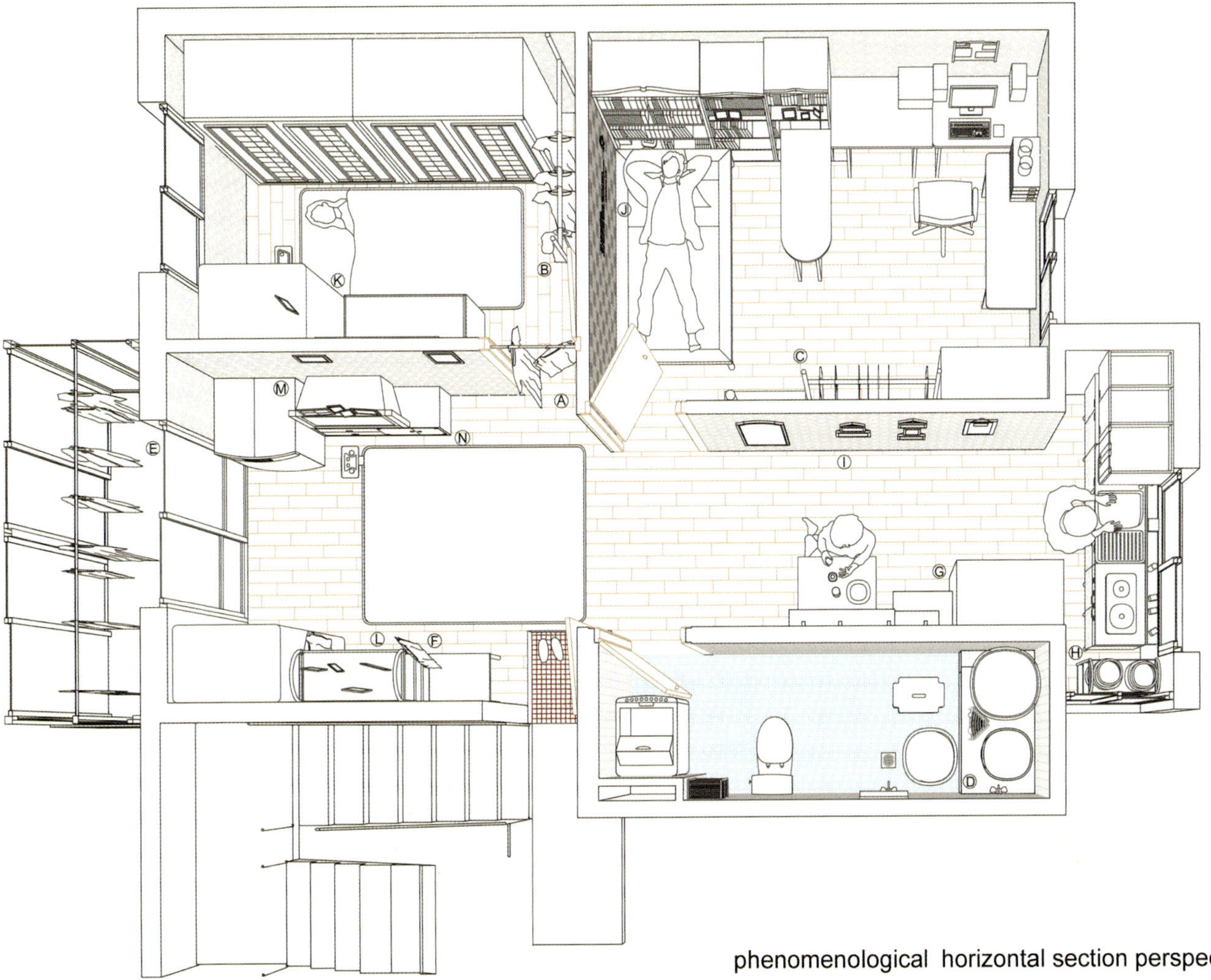

phenomenological horizontal section perspective

phenomenological vertical section perspective

Do not terrified;
do not discouraged, for de LORD
your God with be you Wherever you go
- Joshua: 1:9 -

Hugo Coloma Garcia

A home is not just a house...

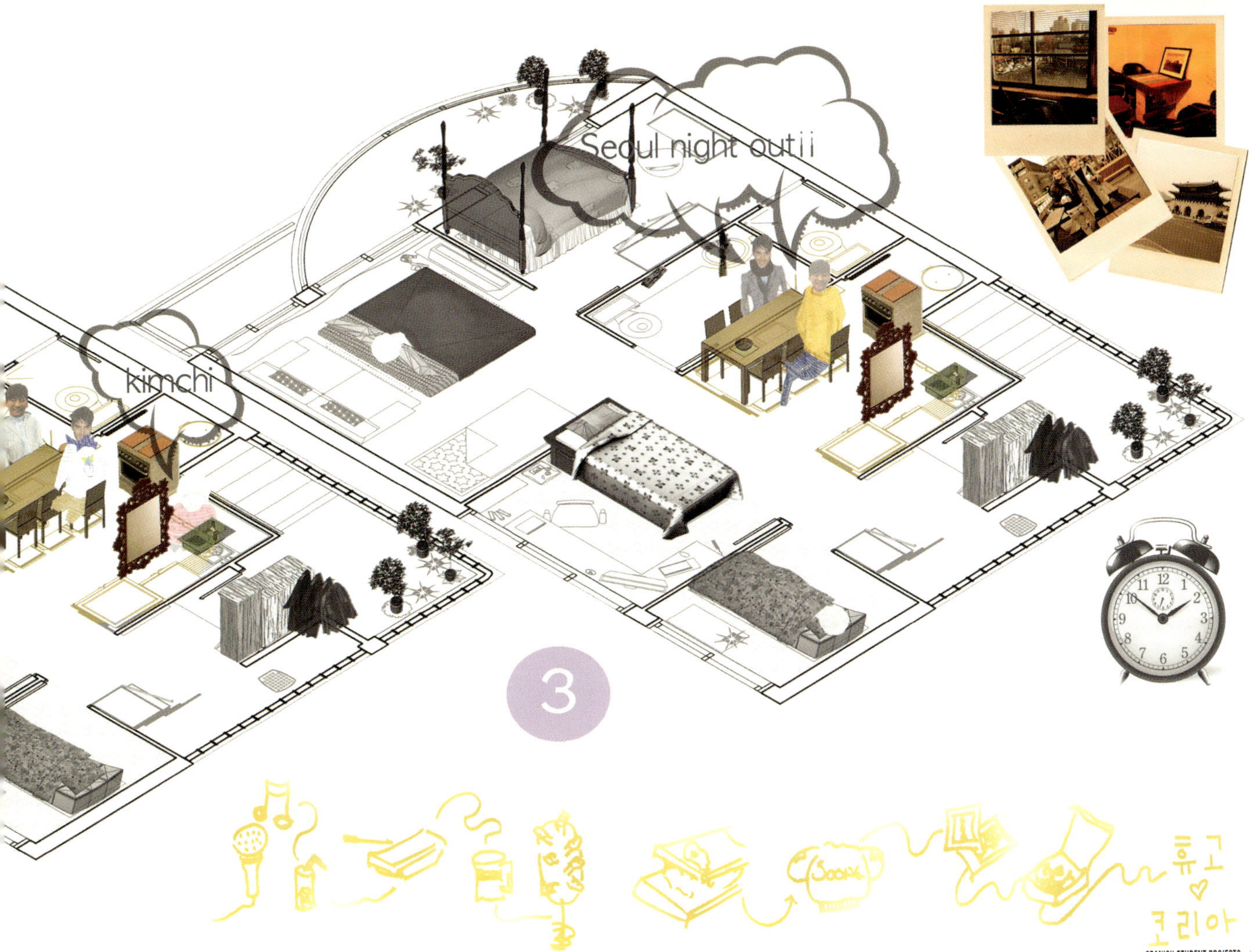
Seoul night out!!
kimchi
3
휴고
코리아

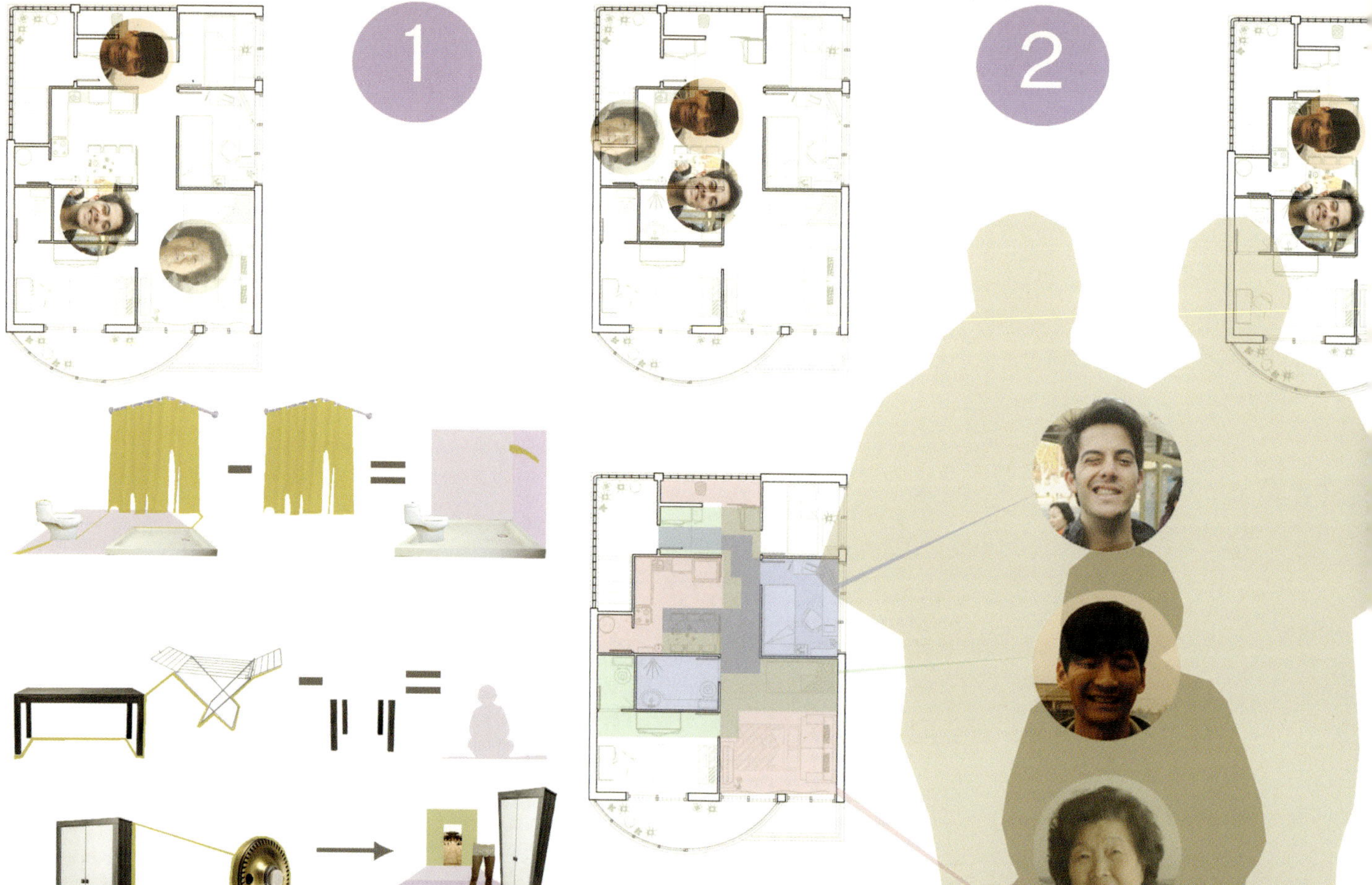
1
2

3

Be my citizen...

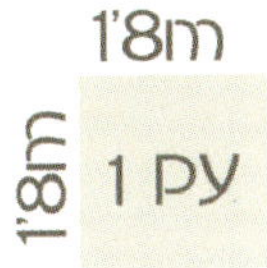

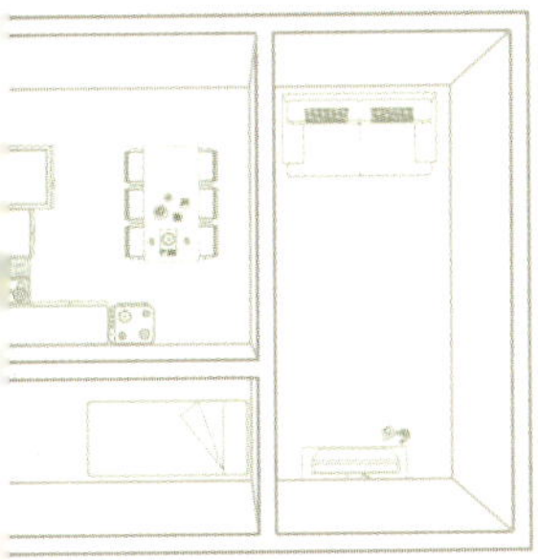

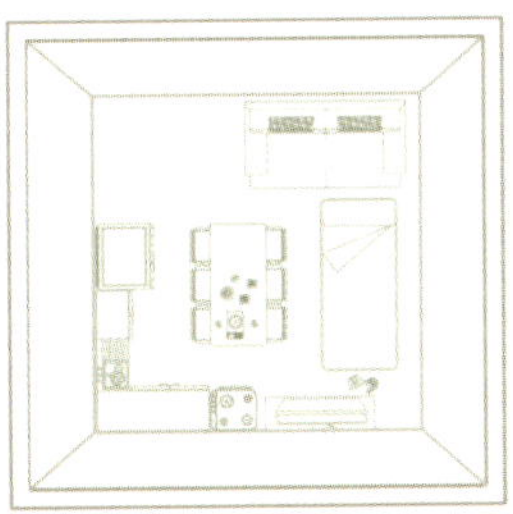

My experience took place in the domestic sphere, trying to study and develop ideas from what I saw and live there. I tried to do it in a very personal way, learning from my personal experience as from the place where I lived.

From the beginning I was interested in the parts of the house that were used for a private activity or for a public one. The diferent activities ,that users did, began to mix developing a Joint acivity or a range of activities which used the same place.

The idea of an open house allows different areas to receive different activities. This avoids the idea of a room with only one activity, and even break the boundary between public and private spaces in the house.

The first situation happens the early hours of the morning. Everyday activities take place in the same space. One of the users does daily activities but not up to a normal height, but rather using the ground as the base

Then the bathroom stops to be differented into zones, for all be a wet area where water covers the entire space. So you have to finish your daily activity in the common area.

In the second activity, with a family dinner users share the space doing the same activity.

The last activity showed that after a night out and coming home at daybreak the common space becomes a range of night activities which end with the users sleeping together.

To conclude: "Be my city .. Zen" has two main points. First, share your privacy in a public area of the house begins to transform not only the user, also the house. Because of a house, with different characters using the space differently, must be developed to contain all these situations. While playing around with my attempt to transform into my partner in some moments of the experiences, had helped me to understand the city from the point of view of the citizen, not just as a tourist. Any form of transformism allows a modification that can be fast and reversible. This idea was developed from the difficult distinction between public and private spaces inside the house.

AIRSTAR
AVENUE
Winter
SALE

"... difference enrich the living and personal development"

"What surprised me the most was the fact I could not recognize a historical centre in Seoul with a recognizable pattern... everything was the same"

Javier Esquiva

"... this was one of my first real architectural experiences"

"I would bring to Alicante the 24h open city"

Adrián Beltrán

"The idea we had about architecture has been modified by this experience"

"Seoul = density + contrasts"

"I would bring the Jimjilbang as it is the most democratic space I have ever been.

We were all the same"

"We architects have the need of knowing and this workshop has given us the chance to do so"

Eduardo Landete

Firstly, the theme of this workshop, 'Be My City', was really interesting to me. We were a kind of strangers in Alicante. And Spain students were the same. I was constantly trying to interpret city Alicante from my own experience. I was able to interpret city 'Seoul' trough my Spainish students' lens.

Jihyun Park

During the workshop, we were able to understand each other in a very short time.

And we were able to learn a flexible new approach for architectural thinking.

Juhye Kim

The way of understanding foreign city is from walking with your local friends.

I realized if we don't admit cultural diversity, then we are loosing the architectural potential.

This 'Korea-Spain exchange' changed my architectural thinking totally.

"Urban situations were originated from people's behavior."

Seonghak Cho

Editor. Seonghak Cho, Juhye Kim, Jihyun Park

"... difference enrich the living and personal development"
"What surprised me the most was the fact I could not recognize a historical centre in Seoul with a recognizable pattern... everything was the same"
Javier Esquiva

"... this was one of my first real architectural experiences"
"I would bring to Alicante the 24h open city"
Adrián Beltrán

"The idea we had about architecture has been modified by this experience"
"Seoul = density + contrasts"
"I would bring the Jimjilbang as it is the most democratic space I have ever been. We were all the same"
"We architects have the need of knowing and this workshop has given us the chance to do so"
Eduardo Landete

Firstly, the theme of this workshop, 'Be My City', was really interesting to me. We were a kind of strangers in Alicante. And Spain students were the same. I was constantly trying to interpret city Alicante from my own experience. I was able to interpret city 'Seoul' trough my Spainish students' lens.
Jihyun Park

During the workshop, we were able to understand each other in a very short time.
And we were able to learn a flexible new approach for architectural thinking.
Juhye Kim

The way of understanding foreign city is from walking with your local friends.
I realized if we don't admit cultural diversity, then we are loosing the architectural potential.
This 'Korea-Spain exchange' changed my architectural thinking totally.
"Urban situations were originated from people's behavior."
Seonghak Cho

Editor. Seonghak Cho, Juhye Kim, Jihyun Park